Secure and Trustworthy Cyberphysical Microfluidic Biochips

Jack Tang • Mohamed Ibrahim
Krishnendu Chakrabarty • Ramesh Karri

Secure and Trustworthy Cyberphysical Microfluidic Biochips

A practical guide to cutting-edge design techniques for implementing secure and trustworthy cyberphysical microfluidic biochips

 Springer

Jack Tang
New York University
Brooklyn, NY, USA

Krishnendu Chakrabarty
Department of ECE
Duke University
Durham, NC, USA

Mohamed Ibrahim
Intel (United States)
Santa Clara, CA, USA

Ramesh Karri
New York University
Brooklyn, NY, USA

ISBN 978-3-030-18165-9   ISBN 978-3-030-18163-5   (eBook)
https://doi.org/10.1007/978-3-030-18163-5

This Springer imprint is published by the registered company Springer Nature Switzerland AG.
The registered company address is: Gewerbestrasse 11, 6330 Cham, Switzerland

To Amy and Simone,
for showing me what truly matters.
— Jack Tang

Preface

Security, trust, and privacy issues arising out of the mass adoption of modern technologies are now commonplace. The rate at which these issues have gone from afterthought to frequently appearing front page news is disconcerting. Interestingly, even though security is now a marketing buzzword, the actual adoption and implementation of security measures in certain industries is still inadequate. When the Internet-of-Things (IoT) movement began to take shape during the 2000s, numerous articles on the security of these highly connected, resource-constrained devices were published. Despite this knowledge, IoT manufacturers proceeded to develop insecure devices due to either cost, time, lack of interest, or poor management decisions. The success of the Mirai botnet in 2017, which leveraged insecure password defaults to take over devices such as DVRs and webcams, is a case in point.

This book concerns the security and trust of cyberphysical microfluidic biochips (CPMBs), with emphasis on the analysis and design of hardware that is robust against tampering. We argue that it is the responsibility of a system designer to anticipate security issues, and we demonstrate this through the development of several hardware-based countermeasures. The work in this book is not reactionary—it is not a series of Band-Aid fixes for systems designed without security implications in mind, which is essentially the practice used by Internet-based services and devices today. Instead, this work is completely anticipatory in nature, building from the ground up with performance under malicious activity as a key design metric.

Security and trust is a broad and sometimes difficult-to-define topic of academic inquiry. This book's definition of security and trust is inspired to a large extent from the field of hardware security, which has historically been understood to refer to semiconductor devices, processes, and their manufacturing models. It includes a large body of work on intellectual property (IP) protection schemes, driven by the interests of large semiconductor and media companies. But it also includes attacks and countermeasures that more directly affect the end user as well as interesting new physical primitives such as physical unclonable functions (PUFs). The field has roots in the VLSI design-for-test community and often utilizes the same tools such as Boolean satisfiability, automatic test program generation (ATPG), and integer linear programming (ILP).

Another commonality hardware security shares with DfT is that of intense initial resistance from industry. The concept of a scan chain was at one point considered to be unnecessary overhead that could be better used for implementing functionality but eventually gained widespread adoption. Similarly, the concept of using a scan chain for attacks and having to design a secure scan chain was met skepticism. As such, security researchers often face a catch-22 when developing any new work. This book takes the stance that it is better to anticipate security threats rather than wait for them to appear. To that end, this book features three design techniques for the prevention, detection, and mitigation of actuation tampering attacks on CPMBs.

For prevention, we present the concept of a tamper-resistant pin-constrained digital microfluidic biochip (DMFB), which manipulates fluids in discrete droplets on a grid of patterned electrodes. The most straightforward implementation of a DMFB brings out each electrode to a pin which is then connected to driver circuitry. To overcome the high pin count, pin-constrained DMFBs short compatible electrodes to the same pin. This trades off the pin count for a reduction in freedom of droplet movement and forms the basis of the tamper resistance property, which we then optimize for using graph theory and integer linear programming.

We then propose a randomized checkpoint system for the purposes of detecting an actuation tampering attack. Practical DMFBs are often integrated with sensors for the purpose of monitoring the progression of assays since they are prone to many failure modes. Due to cost and resource constraints, these sensing systems are limited in the number of locations on the biochip that can be monitored. By randomizing the inspection of electrodes in time and space, we accommodate these constraints while making it probabilistically difficult for an attack to evade detection.

Finally, we develop a design framework for tamper-mitigating microfluidic routing fabrics. Such reconfigurable primitives are attractive to end users for their convenience, but they also present an opportunity for an attacker to force the biochip into an unintended state. This book describes a method for analysis and synthesis of routing fabrics that probabilistically mitigate fault injection-based tampering attacks.

The work featured in this book fits into the broader context of hardware security as a new subfield, which we call *cyberphysical microfluidic biochip security* or CPMB security. The implementation of CPMB security measures shares many similarities with those used in other cyberphysical systems and electronic devices, but the key differentiator is that we leverage unique microfluidic aspects of these systems wherever possible. The authors hope that this book functions as a resource and inspiration for the designers of next-generation microfluidic systems.

The authors acknowledge the financial support received from the Army Research Office under grant number W911NF-17-1-0320, the National Science Foundation under grant CNS-1833624, the NYU Center for Cyber Security, and the NYU-AD Center for Cyber Security.

Brooklyn, NY, USA   Jack Tang
Santa Clara, CA, USA   Mohamed Ibrahim
Durham, NC, USA   Krishnendu Chakrabarty
Brooklyn, NY, USA   Ramesh Karri

Contents

Chapter 1
Cyberphysical Microfluidic Biochips

1.1 Introduction

A cyberphysical microfluidic biochip (CPMB) is, at its core, simply a device for the manipulation of fluids. The term *microfluidic* refers to the fact that the fluid scales involved are microliters or less. More broadly, microfluidics refers to a vast number of technologies used to realize such small scale fluid manipulation. These technologies are necessarily interdisciplinary in nature, spanning the fields of material science, physics, electrical engineering, biochemistry, and computer-aided design, to name a few. The substrate upon which microfluidic fluid manipulation occurs is often called a *microfluidic biochip* or biochip. *Chip* borrows from integrated circuits while *bio* references the application in biochemical sciences. The term *cyberphysical* refers to the integration of sensors, actuators, and intelligent control to significantly expand the capabilities of the microfluidic biochip. All of this together can be called a cyberphysical microfluidic system, or platform, which is the terminology we will use on occasion if the context is appropriate. In general, however, we will use CPMB to refer to the entire system. With all these components in place, one should be well-positioned to realize the ideal device: a compact, user-friendly platform for performing laboratory protocols. Thus, other terms often used interchangeably with microfluidics include: micro-total-analysis-systems (μTAS), lab-on-a-chip (LOC), and sample-to-answer systems.

In this chapter, we review the technologies used to realize a CPMB. While there are countless ways to manipulate small volumes of fluids, in this book we concentrate on two design paradigms that are amenable to cyberphysical integration and design automation techniques: digital microfluidic biochips (DMFBs) and flow-based microfluidic biochips (FMFBs). As we will see in the following chapter, the adoption of such design techniques leads to questions about security and trust. The background information presented here will be assumed throughout the remainder of the book, though important details will be reviewed where necessary.

© Springer Nature Switzerland AG 2020
J. Tang et al., *Secure and Trustworthy Cyberphysical Microfluidic Biochips*,
https://doi.org/10.1007/978-3-030-18163-5_1

1.2 Microfluidics

The literal definition of microfluidics refers only to the scale of the fluids being manipulated. Therefore, the term is ambiguous with regard to implementation details as a multitude of technologies can perform fluid manipulation at small scale. The development of microfluidics has been motivated by the seemingly unlimited benefits afforded by scaling traditional laboratory procedures into the sub-microliter domain:

1. *Reduced sample and reagent usage*: Many applications analyze or process samples, combining them with various reagents. Obviously, scaling down to smaller volumes necessarily reduces fluid consumption. This is attractive because reagents can be costly, while samples can be difficult to obtain and sometimes even harder to replace.
2. *Increased performance*: A high surface area-to-volume ratio means that temperature can be better controlled, leading to increased performance in resolution and sensitivity [1].
3. *Low power consumption*: Low mass means less force is required to move fluids. Low power consumption benefits devices designed to be portable or operated in remote settings. It also permits tighter physical system integration, as it implies less heating in battery and power conversion systems. Excess heating can also potentially interfere with assay operation.
4. *Portability*: Another natural by-product of small scales. Unfortunately, the bottleneck for portability has turned out to be the associated support hardware required to operate a microfluidic biochip [2].
5. *Physical effects*: Fluids undergo laminar flow rather than turbulent flow at small scales. This works to the microfluidic biochip designer's advantage, as fluid flows can be more easily predicted and precisely controlled. Surface/interfacial tension and capillary forces are dominant at microscale, and can be leveraged to move fluids through channels [3].
6. *Automation*: Typical laboratory procedures require extensive human intervention in the workflow, involving loading of samples and reagents and moving the fluids between various instruments. Scaling permits the integration of several disparate system components on one substrate, eliminating the need for human intervention while also permitting the implementation of more complex protocols.

For an interesting perspective on the development of microfluidics, the reader is referred to [4], which details how molecular analysis, biodefense, molecular biology, and microelectronics each contributed to microfluidics as we know it today.

Fluid manipulation mechanisms can vary significantly; examples of technologies not covered in this work include paper microfluidics [5], droplet microfluidics [6], and acoustic actuation [7]. Despite all this variety, we can still discuss microfluidics in categorical terms, which is the topic of the remainder of this section.

1.2.1   Self-Contained Microfluidic Biochips

At the time of this writing, there exists a significant gap between the idealized form of a microfluidic device and its real-world implementation. The ideal can be imagined to take the form of the Star Trek medical tricorder: handheld, compact, non-invasive, able to both scan a sample and produce the result with fast turnaround times. The reality is that these devices often exist as clunky benchtop laboratory platforms. Some implementations even exist as a kludge requiring various external laboratory instruments to function properly. This clearly limits their application as true lab-on-chip platforms. Devices that do not have this limitation are *self-contained*, and can be classified according to their actuation mechanism [8].

- *Passive* self-contained microfluidic biochips use mechanisms such as capillary flow and colorimetric detection to provide functionality that does not depend on any external support. Paper microfluidics and home pregnancy tests are examples of passive microfluidics, and have had tremendous commercial success due to their simplicity and accuracy.
- *Hand-powered* systems require human action to provide the driving force, whether through pressing a syringe, pipetting, or squeezing blister packs.
- *Active* systems use electronics, sensors, actuators, pumps, and control valves to automate the processing of fluids.

Practical systems may have hybrid characteristics; for instance, many commercial systems still rely on humans to pipette the samples and reagents into a cartridge, which is then loaded into an automated processing unit. This book focuses exclusively on active microfluidic biochips. In some cases, we will analyze them in terms of their implementation as they often exist today, but in others we will project toward future implementation as fully self-contained systems.

1.2.2   Cyberphysical Integration

A biochip only contains the substrate upon which fluids are manipulated. For some devices, such as those based on capillary actuation, this suffices to achieve the required functionality. But for the implementation of more complex protocols on DMFBs and FMFBs, cyberphysical integration is required: the use of sensors, actuators, and intelligent control to influence physical actions.

Sensor feedback and control can work together to implement real-time branching decisions in conditional protocols [9], or can choose to take corrective actions in the event of a hardware fault or an error such as incomplete mixing [10]. Cyberphysical integration also permits a CPMB to communicate to other devices, upload collected data for analysis, and perform in-field updates. This opens up the possibility of security vulnerabilities and is the topic of the next chapter.

Examples of sensors employed in CPMBs include: fluorescence detectors, cameras, and sometimes proprietary designs such as Oxford's Nanopore technology which measures changes in electrical current as molecules pass through a nano-sized hole. In some cases, sensors can be integrated with the biochip. But more typically, these components, as well as the controller and the interface for loading of fluids onto the biochip, require devices be integrated beyond chip scale. This is due to the nature of common methods used to fabricate biochips; while early attempts at microfluidics borrowed from silicon fabrication, today, polydimethylsiloxane (PDMS) and even low-cost printed circuit board (PCB) technology are frequently employed.

1.2.3 Computer-Aided Design

Increasingly sophisticated fluid manipulation techniques and protocols have necessitated the development of computer-aided design methodologies, inspired in part by the great strides made by the VLSI design community [11]. Some of these innovations are borne out of the monotony of manual design and operation, while others have been developed to enable high-throughput, high complexity devices. The literature of CAD for microfluidics is too vast to describe here and likely beyond the scope of this book, but we do highlight some important contributions for the interested reader.

- *Architectural synthesis*: The problem of constructing an application-specific biochip architecture that satisfies certain reliability goals can be tackled using metaheuristics such as tabu search and simulated annealing [12].
- *Reliability and fault-tolerance*: Microfluidic devices are plagued by a number of failure modes. Some of these can be induced quickly during the course of normal usage. By optimizing for reliability during the design phase, significant improvements in yield and product lifetime can be achieved [13, 14].
- *Test*: The challenge for efficient testing in microfluidics is unique from VLSI test in that, while the complexity is often smaller, the time constants are longer and the process of testing itself can degrade reliability. Approaches to test include modeling the microfluidic biochip as an equivalent set of logic gates and applying algorithms from VLSI test, with some minor modifications [15–18].
- *Protocol optimization*: While we often make the assumption that a biochemist will provide a ready-made protocol for implementation on a platform, in many cases this process can be automated and optimized. For example, sample preparation is the process of diluting and/or mixing a sample fluid into several known concentrations. Dilution and mixing, if not done properly, can lead to excess fluid consumption and degradation of the biochip [19–22].

1.2.4 Design Flows

Currently, fabrication of CPMBs is vertically integrated; the systems are designed and fabricated within the same laboratory or factory. However, it is reasonable to expect that as the technologies become more advanced, a horizontal supply chain similar to the semiconductor industry will be adopted. This will also permit the democratization of biochip design, as parties interested in developing novel biochemical protocols will no longer have to contend with bringing up their own fabrication facilities.

A microfluidic biochip can follow one of the two design flows illustrated in Fig. 1.1. In the general-purpose flow, first the functional specification of what assay the device is intended to perform is described by the *biocoder*. The biocoder produces a high-level specification using a language such as BioCoder [23], which formalizes the assay description with standardized operations such as dispensing, mixing, heating, and detection. The high-level specification is then passed along to the *biochip designer*, who is responsible for aggregating the hardware and software components used to build the biochip. The designer will use hardware sourced from a biochip vendor and software tools from a CAD tool vendor. The hardware vendor provides information on manufacturing capabilities, such as design rules for etching electrodes, channels, valves, and pumps and fluids that can safely be manipulated on the platform. The hardware vendor fulfills the same role as the foundry in an IC design flow, and may distribute a process design kit to aid the biochip designer.

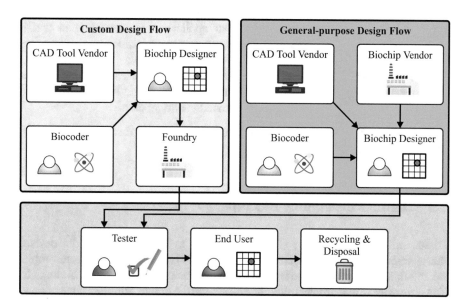

Fig. 1.1 Microfluidic biochip design flows can be either custom or general-purpose. Each stage in the design flow can introduce new security vulnerabilities

The CAD tool vendor writes synthesis software that converts the assay specification into instructions executable on the biochip hardware. After the design is fabricated and integrated, it is sent to the *tester* for validation, and then finally the *end user*. Once the device has exceeded its useful lifetime, it must be collected for *recycling and disposal* at a facility that accepts biochemical waste. An application-specific biochip design flow is similar, but design information must be synthesized by the biochip designer before being given to the foundry for fabrication.

1.2.5 Applications

Some of the major applications for microfluidics include:

- *Scientific research*: This includes activities such as drug discovery and various "omics" including genomics and proteomics. Historically, such activities were limited to laboratories with significant investment in equipment and support. With the adoption of microfluidics, high-throughput sequencing can be achieved at facilities with limited resources.
- *Defense and public safety*: Microfluidic devices can be engineered to detect hazardous chemicals and pathogens. In fact, this application was one of the major motivations for the development of microfluidic technology [4]. Examples of microfluidic devices engineered for public safety applications include monitoring of pollutants in the environment [24, 25], detection of pathogens or antibiotics for food safety assurance [26], and detection of chemical and biological weapons [27]. This particular application is interesting in that it provides a measure of physical security, but can be undermined by an insecure implementation.
- *Medical diagnostics*: The ability to rapidly diagnose diseases and multiplex assays, especially in remote settings outside of clinics and hospitals, will be a boon for patient empowerment and improved quality of care.

We will revisit many of these applications in the context of security and trust in the following chapter.

1.3 Digital Microfluidic Biochips

Digital microfluidic biochips (DMFBs) refer to a class of microfluidic devices that manipulates fluids in discrete droplets [28, 29]. This is distinct from droplet microfluidics [6], which are devices that encapsulate target molecules in droplets. The difference is that droplet microfluidics shuttle the droplets through channels and chambers, more analogously to flow-based microfluidics, while digital microfluidic biochips move in discrete distances. The term DMFB does not presuppose any

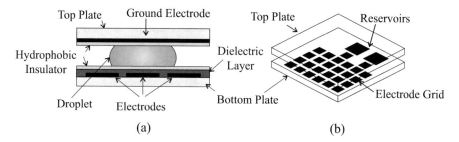

Fig. 1.2 Structure of a digital microfluidic biochip. (**a**) Side view. Droplets can be manipulated through the application of suitable control voltages. (**b**) High-level view. A general-purpose DMFB typically consists of a patterned grid of electrodes

particular physical actuation method; however, over the last decade it has become synonymous with electrowetting-on-dielectric (EWOD) [30].

The structure of a typical DMFB consists of a patterned grid of electrodes that is coated with a hydrophobic layer (Fig. 1.2). The substrate can also be patterned to form reservoirs and channels, and the electrode grid can be either rectangular/square or irregular shape. If there is a top plate present, it is often fabricated with a translucent conductive material in order to observe droplet movements and electrically ground the plate. Application of a voltage to the control electrodes induces the EWOD on the droplets, which can be harnessed for fluid handling operations. Use of a top plate prevents evaporation of fluids, and can decrease the required operating voltage by allowing droplets to move about in a viscous medium such as silicon oil.

1.3.1 Electrowetting-on-Dielectric

Electrowetting-on-dielectric can be modeled using the Young–Lippmann equation [28]:

$$\cos \theta = \cos \theta_0 + \frac{\epsilon_0 \epsilon_r V^2}{2 \gamma d} \tag{1.1}$$

where θ is the static contact angle between the droplet and the substrate, θ_0 is the static contact angle without voltage, ϵ_r is the relative permittivity of the dielectric, V is the voltage, γ is the surface tensions of the filler media, and d is the dielectric thickness. By varying the applied voltage and controlling the fabrication of the DMFB, the contact angle can be modulated and implement operations such as dispensing, transporting, mixing, and spitting of droplets. These operations can then be utilized as part of a complex bioassay for applications such as DNA amplification and in-vitro protein synthesis [28, 31].

In practice, system designers typically abstract away the underlying physics of EWOD droplet movement unless some aspect of it directly impacts performance or reliability. Generally it suffices to imagine a grid of electrodes upon which a controller sends high or low voltages. Movement of droplets is induced by applying a low voltage to an electrode with a droplet and a high voltage to an adjacent electrode. Control signals that are used to implement a bioassay are called *actuation sequences*. The control signals are abstracted to logic high (1) and logic low (0), and can be sent to the DMFB from a digital controller unit through proper level-translation circuitry (as DMFBs often require high voltages, sometimes on the order of hundreds of volts!). The rate at which actuation sequences can be applied to the DMFB is limited by the physics of the droplet movement, and in typical systems is on the order of hundreds to several thousand hertz [28].

1.3.2 High-Level Synthesis

The complexity of generating actuation sequences for even relatively simple assays on small DMFB arrays is prohibitive for manual entry. The concept of high-level synthesis was adopted from VLSI early on to cope with this burden [11]. In general, the flow for DMFB high-level synthesis starts with a specification for the bioassay to be executed on-chip. This specification is written in a high-level descriptive language and passed to the synthesis software for processing. Synthesis typically consists of solving three NP-hard optimization problems (Fig. 1.3):

1. *Scheduling* is the determination of the order of operations such that precedence and resource constraints are met [11].
2. *Placement* is the allocation of physical biochip resources for the implementation of operations such as mixing [13]. Two methods are commonly employed: free placement, utilizing algorithms adopted from FPGA floor planning [32], and

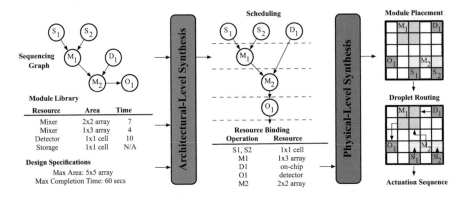

Fig. 1.3 High-level synthesis design flow for digital microfluidic biochips

virtual topology placement [33, 34], which assumes that only certain areas on the chip can be used for operations. Virtual topology often leads to higher quality results in terms of compactness and routability.

3. *Routing* is the determination of droplet transport paths from their dispense ports to operating modules, and finally to an output or waste port. Algorithms from VLSI routing can be adopted for this purpose, such as Soukup's maze router [35], and routing for multiple droplets can be compacted such that they occur simultaneously [36].

The optimization objective is typically completion time, but reliability and testability are also used [11, 37]. Each of these phases bears similarity to their VLSI counterparts, but differs in that the resources used for processing and routing in a DMFB are the same. That is, in VLSI synthesis, transistors and gates are used to implement logic while wires are used for interconnection. In DMFBs, electrodes are used both to process fluids and to route them. Alternate synthesis flows have been proposed that tackle all phases simultaneously, which can often lead to higher quality optimization results [17, 38, 39].

1.3.3 Checkpoint-Based Error Recovery

DMFBs have been shown to be susceptible to a wide array of faults, attributed to the large variety of potential defects such as residual charge and protein adsorption during bioassay execution [40]. Optimization-based synthesis and checkpoint-based error recovery have been shown to be one of the most effective means of overcoming this fundamental weakness of DMFB technology.

Checkpoint-based error recovery monitors the execution of the DMFB at specific locations in time and space. If an error is detected, resources are diverted to recover in such a way that is more efficient than simply re-executing the assay. Charge-coupled device (CCD) cameras are often utilized to provide the real-time sensor feedback to the biochip controller. CCD cameras are reconfigurable and able to determine the state of the biochip at arbitrary locations and times. Software executing on the controller takes the images captured by the camera and extracts droplet presence, volume, and concentration using pattern-matching algorithms with template images [41, 42]. Proposed approaches to pattern-matching include generating a correlation map for the entire biochip and determining location from electrodes with the highest correlation [41], as well as performing a single droplet correlation after cropping the image at specific locations [10]. While CCD imaging is flexible and effective, it may not be usable for assays that utilize light-sensitive reagents [43]. We also note that other types of sensing hardware can implement checkpoints, as was recently demonstrated for a reliability-hardening technique [44].

1.3.4   Pin-Constrained DMFBs

One of the largest contributors to a DMFB's cost and complexity is the number of pins required to drive it [45]. The earliest general-purpose DMFBs required one IO pin from the driver circuitry for each electrode on-chip, a scheme termed *direct addressing*. This can quickly become impractical, even for modestly sized designs. For instance, a demonstrated immunoassay DMFB targeted for point-of-care testing requires over 1000 pins [46]—which easily exhausts the number of pins on common MCU packages—if direct addressing is used. *Pin-constrained* DMFBs reduce the pin count with a restriction on the droplet degrees of freedom. Pin-constrained DMFBs can be generated as the final step in the high-level synthesis flow, or can be considered in the overall biochip design [47, 48]. The same immunoassay biochip described in [46], when pin-constrained, uses only 64 pins to drive over 1000 electrodes.

1.3.5   Commercialization

At the time of this writing, the authors are aware of only three commercial offerings that utilize digital microfluidic technology, one of which has been discontinued:

- The oldest among these is the discontinued Illumina NeoPrep Library System [49], which was developed using technology acquired from Advanced Liquid Logic, which itself was spun off from research at Duke University. Unfortunately for researchers in digital microfluidics, this platform was seen for a long time as the poster child of successful commercial DMFB platforms, and was frequently referenced in papers.
- The Baebies SEEKER [50] is a DMFB platform for newborn lysosomal disorder screening, which was also spun off from Duke University. This platform is promising as it targets an interesting medical diagnostic application and has already been granted FDA approval.
- The Oxford Nanopore VolTRAX [51] is a device that is still under development, with some prototypes being released to a select group of users. Based on company literature, it appears that this device will be a USB-powered platform for sample preparation applications and integration with Oxford's other technologies. The distinguishing characteristic of this device is its compact form factor, which will truly enable analysis in remote settings.

1.3.6   Open-Source Platforms

Digital microfluidics has a growing open-source presence. There are presently two platforms available for experimentation and one software framework for DMFB synthesis.

- The Auryn.bio OpenDrop [52] platform is a low-cost, DC actuated, PCB-based digital microfluidics platform designed to be easily accessible. It currently has limited support and it is not clear if production will continue.
- The Wheeler DropBot [53] features AC actuation and has several publications detailing its usage in real-world scenarios (including a field trial in Kenya). Early iterations required expensive external laboratory equipment to function, but the platform is being continuously improved and is now at version 3.0. Software is also provided for monitoring and control of the biochip, but it does not implement any advanced automation techniques such as error recovery.
- UC Riverside's MFStaticSim [54] is a C++ framework for synthesizing actuation sequences. It implements several scheduling, placement, routing, pin mapping, and wire routing algorithms, is well-documented, and also includes a useful droplet visualization tool.

1.4   Flow-Based Microfluidic Biochips

Flow-based microfluidic biochips (FBMBs) use etched microchannels, microvalves, and micropumps to direct fluids through networks of reaction chambers [3]. The flow of fluids is continuous, in contrast to digital microfluidics. FBMBs are an effective means of realizing the laboratory-on-a-chip, as they can be densely integrated and fabricated quickly and easily [4, 55–57]. In fact, FBMBs are the foundational technology for the concept of microfluidic large-scale integration (mLSI) [57, 58].

These devices are, in general, not reconfigurable, though the valves do need to be controlled and operated in the proper sequence to achieve the desired functionality. Efforts have been made to develop reconfigurable flow-based biochips (RFBs), increasing their complexity and giving rise to a class of devices that are analogous to field-programmable gate arrays (FPGAs).

1.4.1   Fabrication

Early flow-based microfluidic devices were fabricated using silicon and glass substrates, borrowing techniques from the semiconductor industry [4]. Currently, polydimethylsiloxane (PDMS) is the favored material due to its relatively low cost and ease of fabrication, elasticity, and surface properties [56, 59]. PDMS biochips are fabricated using a process called multilayer soft lithography: the biochip layout is patterned onto silicon wafers by lithography, and are used as molds. A thin layer of PDMS is deposited onto the wafers through spin coating, and subsequently baked to become an elastomer. Holes are punched to define entry and exit points for fluid and pressure sources. Multiple layers can be aligned and bonded together to form a complex network of channels, though more typically only two layers are used.

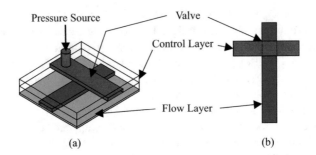

Fig. 1.4 Structure of a flow-based valve biochip. (**a**) A pressure source is connected to the control layer (red), while fluids to be manipulated circulate in the flow layer (blue). (**b**) Intersections define the valves, as pressure in the control layer pinches off the flow layer

1.4.2 Microvalves

One of the fundamental building blocks of FMFBs is the microvalve. Figure 1.4 illustrates the typical construction of a flow-based biochip. Two elastomer layers are fabricated with channels and bonded together. One layer is comprised of flow channels for the manipulation of samples and reagents. Flow is induced by pumps connected at the end of each fluid reservoir. The next layer comprises a network of control channels, which are connected to external pressure sources. Activation of the pressure sources causes a deflection where the control and flow channels intersect, forming a valve. Activation can be designed to be normally-on or normally-off. In a normally-on FMFB, application of pressure in the control layer forces closure of the valve, and movement of the fluid in the flow channel is interrupted.

1.4.3 Fully-Programmable Valve Arrays

Fluid valves and channels can be arranged into intersecting rectangular grids to implement fully-programmable valve arrays (FPVAs) [60]. FPVAs permit reconfiguration into arbitrary flow networks. High density can be achieved to realize microfluidic large scale integration (mLSI) [57]. Mixing can also be achieved through sequencing of the valve states. This can be leveraged for high-throughput processing and parametric studies [57, 58]. For instance, M input samples can be tested against N different reagents, for a total of $M \times N$ unique reactions. Applications for this type of device are numerous and include surface immunoassays and digital polymerase chain reaction [61]. The design of flow control networks for FPVAs can be automated to allow complex protocols and the targeting of design optimizations such as pin-count minimization [62].

1.4.4 Routing Crossbars

Recently, routing crossbars (alternately, routing fabrics) have been proposed to enable rapid prototyping and real-time adjustments to biochemical protocols [60, 63]. Already, this concept has been applied in research on platforms for single-cell analysis and has shown promise as a means for creating more space-efficient systems.

The routing crossbar is based on the concept of a two-input, two-output transposer (Fig. 1.5a), which itself is constructed using a network of channels and valves. In [63], this primitive was constructed using a polydimethylsiloxane (PDMS) substrate with ablated polycarbonate stacked above. The valves are formed as discontinuities in the channels, and can be designed to be normally open or normally closed. An elastomeric membrane covers the valve. This membrane distends into the gap upon vacuum actuation (or pressurized actuation), causing fluid to flow through (or be blocked). An alternative transposer is shown in Fig. 1.5b. One output port can select between two input ports, forming a microfluidic 2-to-1 multiplexer, or alternately, a 1-to-2 demultiplexer. The transposer primitive is then used to build more complex routing fabrics that can select between an arbitrary number of inputs and outputs.

1.4.5 Commercialization

Among the myriad of microfluidic technologies available today, flow-based microfluidics are one of the most developed and have been successfully commercialized in benchtop platforms. One notable example is the Fluidigm BioMark HD [65], which has been on the market since the mid-2010s and can accommodate different types of their proprietary integrated fluidic circuits (IFCs). IFCs are available for performing digital PCR and targeted DNA sequencing in the form of a one-time use cartridge.

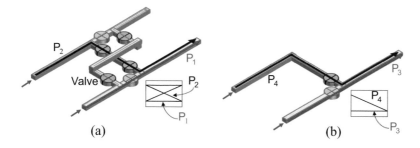

Fig. 1.5 A schematic view of transposers [64]: (**a**) a 2-to-2 transposer primitive. Route P_1 enables the valves in green and disables the valves in grey, causing both fluids to be passed straight through, while route P_2 enables the opposite set of valves to make fluids cross over. (**b**) a 2-to-1 transposer primitive. Enabling either path P_3 or P_4 performs fluid multiplexing

1.5 Summary and Conclusion

In this chapter, we described the basic operating principles of digital and flow-based microfluidics along with their fabrication, design, and progress toward commercialization. All together, these technologies show great promise for realizing the ideal of a microfluidic biochip. Design automation techniques can be successfully applied to microfluidic biochips, greatly reducing design time and enhancing ease of use. This hints at the amenability of these technologies to security and trust, as the remainder of this book will demonstrate.

References

1. S. Pennathur, C. Meinhart, H. Soh, How to exploit the features of microfluidics technology. Lab Chip **8**(1), 20–22 (2008)
2. C.M. Klapperich, Microfluidic diagnostics: time for industry standards. Expert Rev. Med. Devices **6**(3), 211–213 (2014)
3. T.M. Squires, S.R. Quake, Microfluidics: fluid physics at the nanoliter scale. Rev. Mod. Phys. **77**(3), 977 (2005)
4. G.M. Whitesides, The origins and the future of microfluidics. Nature **442**(7101), 368–373 (2006)
5. Y. Xia, J. Si, Z. Li, Fabrication techniques for microfluidic paper-based analytical devices and their applications for biological testing: a review. Biosens. Bioelectron. **77**, 774–789 (2016)
6. S.-Y. Teh, R. Lin, L.-H. Hung, A.P. Lee, Droplet microfluidics. Lab Chip **8**(2), 198–220 (2008)
7. B. Hadimioglu, R. Stearns, R. Ellson, Moving liquids with sound: the physics of acoustic droplet ejection for robust laboratory automation in life sciences. J. Lab. Autom. **21**(1), 4–18 (2016)
8. M. Boyd-Moss, S. Baratchi, M. Di Venere, K. Khoshmanesh, Self-contained microfluidic systems: a review. Lab Chip **16**(17), 3177–3192 (2016)
9. D. Grissom, C. Curtis, P. Brisk, Interpreting assays with control flow on digital microfluidic biochips. ACM J. Emerg. Technol. Comput. Syst. **10**(3), 24 (2014)
10. Y. Luo, K. Chakrabarty, T.-Y. Ho, Error recovery in cyberphysical digital microfluidic biochips. IEEE Trans. Comput. Aided Des. Integr. Circuits Syst. **32**(1), 59–72 (2013)
11. F. Su, K. Chakrabarty, High-level synthesis of digital microfluidic biochips. ACM J. Emerg. Technol. Comput. Syst. **3**(4), 1 (2008)
12. M. Alistar, P. Pop, J. Madsen, Synthesis of application-specific fault-tolerant digital microfluidic biochip architectures. IEEE Trans. Comput. Aided Des. Integr. Circuits Syst. **35**(5), 764–777 (2016)
13. F. Su, K. Chakrabarty, Module placement for fault-tolerant microfluidics-based biochips. ACM Trans. Des. Autom. Electron. Syst. **11**(3), 682–710 (2006)
14. P. Pop, M. Alistar, E. Stuart, J. Madsen, Design methodology for digital microfluidic biochips, in *Fault-Tolerant Digital Microfluidic Biochips: Compilation and Synthesis* (Springer, Cham, 2016), pp. 13–28
15. T. Xu, K. Chakrabarty, Parallel scan-like test and multiple-defect diagnosis for digital microfluidic biochips. IEEE Trans. Biomed. Circuits Syst. **1**(2), 148–158 (2007)
16. T.A. Dinh, S. Yamashita, T.-Y. Ho, K. Chakrabarty, A general testing method for digital microfluidic biochips under physical constraints, in *Proceedings of IEEE International Test Conference*, 2015, pp. 1–8

17. C.C.-Y. Lin, Y.-W. Chang, ILP-based pin-count aware design methodology for microfluidic biochips. IEEE Trans. Comput. Aided Des. Integr. Circuits Syst. **29**(9), 1315–1327 (2010)
18. K. Hu, F. Yu, T.-Y. Ho, K. Chakrabarty, Testing of flow-based microfluidic biochips: fault modeling, test generation, and experimental demonstration. IEEE Trans. Comput. Aided Des. Integr. Circuits Syst. **33**(10), 1463–1475 (2014)
19. S. Bhattacharjee, A. Banerjee, B.B. Bhattacharya, Sample preparation with multiple dilutions on digital microfluidic biochips. IET Comput. Digit. Tech. **8**(1), 49–58 (2014)
20. S. Bhattacharjee, S. Poddar, S. Roy, J.-D. Huang, B.B. Bhattacharya, Dilution and mixing algorithms for flow-based microfluidic biochips. IEEE Trans. Comput. Aided Des. Integr. Circuits Syst. **36**(4), 614–627 (2017)
21. D. Mitra, S. Roy, S. Bhattacharjee, K. Chakrabarty, B.B. Bhattacharya, On-chip sample preparation for multiple targets using digital microfluidics. IEEE Trans. Comput. Aided Des. Integr. Circuits Syst. **33**(8), 1131–1144 (2014)
22. S. Roy, B.B. Bhattacharya, S. Ghoshal, K. Chakrabarty, Low-cost dilution engine for sample preparation in digital microfluidic biochips, in *Proceedings of the International Symposium on Electronic System Design*, 2012, pp. 203–207
23. V. Ananthanarayanan, W. Thies, Biocoder: a programming language for standardizing and automating biology protocols. J. Biol. Eng. **4**(1), 1 (2010)
24. G. Chen, Y. Lin, J. Wang, Monitoring environmental pollutants by microchip capillary electrophoresis with electrochemical detection. Talanta **68**(3), 497–503 (2006)
25. J.C. Jokerst, J.M. Emory, C.S. Henry, Advances in microfluidics for environmental analysis. Analyst **137**(1), 24–34 (2012)
26. S. Neethirajan, I. Kobayashi, M. Nakajima, D. Wu, S. Nandagopal, F. Lin, Microfluidics for food, agriculture and biosystems industries. Lab Chip **11**(9), 1574–1586 (2011)
27. J. Wang, Microchip devices for detecting terrorist weapons. Anal. Chim. Acta **507**(1), 3–10 (2004)
28. K. Choi, A.H. Ng, R. Fobel, A.R. Wheeler, Digital microfluidics. Annu. Rev. Anal. Chem. **5**(1), 413–440 (2012)
29. R.B. Fair, Digital microfluidics: is a true lab-on-a-chip possible? Microfluid. Nanofluid. **3**(3), 245–281 (2007)
30. M. Pollack, A. Shenderov, R. Fair, Electrowetting-based actuation of droplets for integrated microfluidics. Lab Chip **2**(2), 96–101 (2002)
31. H.-H. Shen, S.-K. Fan, C.-J. Kim, D.-J. Yao, EWOD microfluidic systems for biomedical applications. Microfluid. Nanofluid. **16**(5), 965–987 (2014)
32. K. Bazargan, R. Kastner, M. Sarrafzadeh, et al., Fast template placement for reconfigurable computing systems. IEEE Des. Test Comput. **17**(1), 68–83 (2000)
33. D. Grissom, P. Brisk, Fast online synthesis of generally programmable digital microfluidic biochips, in *Proceedings of the IEEE/ACM/IFIP International Conference on Hardware/Software Codesign and System Synthesis*, 2012, pp. 413–422
34. D.T. Grissom, P. Brisk, Fast online synthesis of digital microfluidic biochips. IEEE Trans. Comput. Aided Des. Integr. Circuits Syst. **33**(3), 356–369 (2014)
35. J. Soukup, Fast maze router, in *Proceedings of the IEEE/ACM Design Automation Conference*, 1978, pp. 100–102
36. T.-W. Huang, T.-Y. Ho, A fast routability- and performance-driven droplet routing algorithm for digital microfluidic biochips, in *Proceedings of the IEEE International Conference on Computer Design*, 2009, pp. 445–450
37. K. Chakrabarty, Design automation and test solutions for digital microfluidic biochips. IEEE Trans. Circuits Syst. I **57**(1), 4–17 (2010)
38. T. Xu, K. Chakrabarty, Integrated droplet routing and defect tolerance in the synthesis of digital microfluidic biochips. ACM J. Emerg. Technol. Comput. Syst. **4**(3), 11 (2008)
39. P.-H. Yuh, C.-L. Yang, Y.-W. Chang, Bioroute: a network-flow-based routing algorithm for the synthesis of digital microfluidic biochips. IEEE Trans. Comput. Aided Des. Integr. Circuits Syst. **27**(11), 1928–1941 (2008)

40. T. Xu, K. Chakrabarty, Fault modeling and functional test methods for digital microfluidic biochips. IEEE Trans. Biomed. Circuits Syst. **3**(4), 241–253 (2009)
41. Y.-J. Shin, et al., Machine vision for digital microfluidics. Rev. Sci. Instrum. **81**(1), 014302 (2010)
42. C.-L. Sotiropoulou, L. Voudouris, C. Gentsos, A.M. Demiris, N. Vassiliadis, S. Nikolaidis, Real-time machine vision FPGA implementation for microfluidic monitoring on lab-on-chips. IEEE Trans. Biomed. Circuits Syst. **8**(2), 268–277 (2014)
43. D. Witters, K. Knez, F. Ceyssens, R. Puers, J. Lammertyn, Digital microfluidics-enabled single-molecule detection by printing and sealing single magnetic beads in femtoliter droplets. Lab Chip **13**(11), 2047–2054 (2013)
44. G.-R. Lu, G.-M. Huang, A. Banerjee, B.B. Bhattacharya, T.-Y. Ho, H.-M. Chen, On reliability hardening in cyber-physical digital-microfluidic biochips, in *Proceedings of the Asia and South Pacific Design Automation Conference*, 2017, pp. 518–523
45. D.T. Grissom, J. McDaniel, P. Brisk, A low-cost field-programmable pin-constrained digital microfluidic biochip. IEEE Trans. Comput. Aided Des. Integr. Circuits Syst. **33**(11), 1657–1670 (2014)
46. R. Sista, Z. Hua, P. Thwar, A. Sudarsan, V. Srinivasan, A. Eckhardt, M. Pollack, V. Pamula, Development of a digital microfluidic platform for point of care testing. Lab Chip **8**(12), 2091–2104 (2008)
47. Y. Luo, K. Chakrabarty, Design of pin-constrained general-purpose digital microfluidic biochips. IEEE Trans. Comput. Aided Des. Integr. Circuits Syst. **32**(9), 1307–1320 (2013)
48. D. Grissom, P. Brisk, A field-programmable pin-constrained digital microfluidic biochip, in *Proceedings of the IEEE/ACM Design Automation Conference*, 2013, p. 46
49. Illumina, Illumina neoprep library prep system 2016. http://www.illumina.com/systems/neoprep-library-system.html/
50. Baebies, Inc., Baebies SEEKER 2017. http://baebies.com/products/seeker/
51. J. Karow, Oxford nanopore provides update on tech developments at London user meeting 2017. http://www.illumina.com/systems/neoprep-library-system/performance-specifications.html
52. R. Trojok, A. Volpato, M. Alistar, J. Schubert, Auryn: adaptor for general-purpose digital microfluidic biochips, July 2016
53. R. Fobel, C. Fobel, A.R. Wheeler, Dropbot: an open-source digital microfluidic control system with precise control of electrostatic driving force and instantaneous drop velocity measurement. Appl. Phys. Lett. **102**(19), 193513 (2013)
54. D. Grissom, C. Curtis, S. Windh, C. Phung, N. Kumar, Z. Zimmerman, O. Kenneth, J. McDaniel, N. Liao, P. Brisk, An open-source compiler and PCB synthesis tool for digital microfluidic biochips. Integr. VLSI J. **51**, 169–193 (2015)
55. D. Mark, S. Haeberle, G. Roth, F. Von Stetten, R. Zengerle, Microfluidic lab-on-a-chip platforms: requirements, characteristics and applications, in *Microfluidics Based Microsystems* (Springer, Dordrecht, 2010), pp. 305–376
56. M.A. Unger, H.-P. Chou, T. Thorsen, A. Scherer, S.R. Quake, Monolithic microfabricated valves and pumps by multilayer soft lithography. Science **288**(5463), 113–116 (2000)
57. J. Melin, S.R. Quake, Microfluidic large-scale integration: the evolution of design rules for biological automation. Annu. Rev. Biophys. Biomol. Struct. **36**, 213–231 (2007)
58. T. Thorsen, S.J. Maerkl, S.R. Quake, Microfluidic large-scale integration. Science **298**(5593), 580–584 (2002)
59. E.K. Sackmann, A.L. Fulton, D.J. Beebe, The present and future role of microfluidics in biomedical research. Nature **507**(7491), 181–189 (2014)
60. L.M. Fidalgo, S.J. Maerkl, A software-programmable microfluidic device for automated biology. Lab Chip **11**(9), 1612–1619 (2011)
61. S. Dube, J. Qin, R. Ramakrishnan, Mathematical analysis of copy number variation in a DNA sample using digital PCR on a nanofluidic device. PLoS One **3**(8), e2876 (2008)

62. K. Hu, T.A. Dinh, T.-Y. Ho, K. Chakrabarty, Control-layer routing and control-pin minimization for flow-based microfluidic biochips. IEEE Trans. Comput. Aided Des. Integr. Circuits Syst. **36**(1), 55–68 (2017)
63. R. Silva, S. Bhatia, D. Densmore, A reconfigurable continuous-flow fluidic routing fabric using a modular, scalable primitive. Lab Chip **16**(14), 2730–2741 (2016)
64. M. Ibrahim, K. Chakrabarty, U. Schlichtmann, CoSyn: efficient single-cell analysis using a hybrid microfluidic platform, in *Proceedings of the Conference on Design, Automation and Test in Europe*, Lausanne, March 2017
65. L.R. Volpatti, A.K. Yetisen, Commercialization of microfluidic devices. Trends Biotechnol. **32**(7), 347–350 (2014)

Chapter 2
Security and Trust

2.1 Why Security and Trust?

News of security breaches are occurring on a near daily basis. It is tempting to make the argument for security and trust based on its perceived self-evidence. Within the last couple of years, we have observed the exposure of Facebook's involvement in massive data harvesting campaigns [1], the leakage of private data from the credit bureau Equifax [2], and the hijacking of voice-activated personal assistants [3]. So while it is true that any system designer today would do well to consider the security implications of their work, we also highlight other major reasons that CPMB security should be considered:

- *Untrusted supply chains*: Distributed supply chains in modern semiconductor manufacturing lead to a complex network of actors with various capabilities and motivations for malicious activities such as hardware Trojan insertion [4] and intellectual property (IP) piracy [5]. Next-generation commercial microfluidic devices are expected to follow suit as the fabrication processes become more sophisticated, and therefore will be susceptible to the same attacks [6].
- *Cyberphysical integration*: While a boon for implementing robust, reliable systems, cyberphysical integration also poses a significant security risk for several reasons. First, it implies that a digital controller unit is present. Typically this controller is implemented with an off-the-shelf microcontroller or single-board computer and can introduce several new attack surfaces such as network interfaces. Second, the physical aspects of the system can be leveraged by an attacker to great effect; not only is data at risk, but now physical assets are susceptible to alteration and destruction. Interestingly, software vulnerabilities can be exploited to achieve physical attacks, while physical modalities can be leveraged to extract information about the software's operation.
- *Increasing adoption rates*: It has been noted repeatedly in the microfluidics literature that this technology has yet to be commercialized at large scale [7]. Security

© Springer Nature Switzerland AG 2020
J. Tang et al., *Secure and Trustworthy Cyberphysical Microfluidic Biochips*,
https://doi.org/10.1007/978-3-030-18163-5_2

and trust is a feature that can be designed and automated into CPMBs, providing a high level of assurance. This is in contrast to macro-scale laboratories, where training, standards compliance, and auditing are the norm and are subject to human oversight. Security-centric design could finally bring microfluidics from an area of predominantly academic study to practical systems for widespread deployment.

- *Prevention*: The computer industry has a history of adopting security measures after the fact. A response-driven approach ensures that solutions are motivated by real problems, but at great cost to the initially affected parties. Microfluidics is still an emerging technology and has few motivational examples to draw upon. But it is precisely for this reason that microfluidics presents an opportunity, as designing for security and trust can prevent such incidents from occurring in the first place.
- *Amenability*: Microfluidic technologies may be highly amenable to security-centric design. Unique microfluidic properties may permit designers to build inherently secure systems. For example, the complexity of such systems is typically less than that of those in VLSI [8], and time scales in microfluidic systems are often orders of magnitude slower than those in electronics [9]. Furthermore, standards have yet to be published as researchers continue to explore novel design concepts and research basic physics and materials [7]. Therefore, there exists an opportunity to incorporate security into developing standards.

2.2   Taxonomy

Here, we provide a comprehensive taxonomy of attack surfaces, threat models, and motivations that could impact CPMBs. We note that while this taxonomy is theoretical, it is entirely grounded in established principles and examples from other domains. The taxonomy also highlights, as much as possible, aspects unique to microfluidics.

2.2.1   Attack Surfaces

An attack surface is a potential entry point for carrying out an attack. Attack surfaces can be categorized according to the following taxonomy:

- *Indirect physical access*: Physical ports can be used in an unintended manner to carry out an attack. These include exposed USB and Ethernet jacks originally included for maintenance or uploading data, as well as the biochip cartridge loading tray. Additionally, many microfluidic platforms are designed around standard, off-the-shelf embedded computers and have a variety of unused physical ports exposed.

- *Direct physical access*: The goal of many microfluidic device designers is to create a truly portable, self-contained lab-on-a-chip. While this has many practical benefits, it makes the platform physically vulnerable. For instance, a device deployed in remote locations for environmental monitoring would be easily tampered with. In especially vulnerable settings, invasive, direct physical attacks may be possible: fault injection through power or clock glitching and tampering with control lines, samples, or reagents. This also has privacy implications as sensor readings could be sniffed off signal traces.
- *Network access*: Network-enabled cyberphysical integration presents a potent attack surface for malicious adversaries looking to perform remote attacks. Zero-day exploits, hacking tools developed by intelligence agencies, unpatched systems, and social engineering all contribute to the insecurity of networked computers. The fact that CPMBs are often deployed in settings where regular maintenance is infeasible, either due to cost or inaccessibility, compounds the issue. Arguably, the only way to prevent abuse of network interfaces is to remove them, but this would severely limit the usefulness of the CPMB.
- *Wireless access*: Wireless interfaces such as Bluetooth and Zigbee provide a convenient means for transmitting data, especially for applications with smartphone integration. This presents an opportunity for attackers within close proximity, but not necessarily in possession, of a microfluidic platform.
- *Design documentation*: The information used to fabricate a microfluidics platform may consist of schematics for circuitry, protocols for a biochemical assay, or layouts for a biochip. When presented in its raw form, design documentation is easily abused for overbuilding attacks.
- *Standard interfaces*: This refers to an interface that a user is normally expected to interact with, such as a touchscreen or a chassis cutout for viewing assay progression. If not carefully designed, such interfaces may reveal privileged information or permit escalation of privileges. For instance, if user-selectable operating parameters are not protected from invalid or dangerous settings, damage could result to the platform. Or, an attacker may attempt to reverse engineer a protocol by carefully designing a set of fluids that can indicate the order of mixing.
- *Side-channels*: Side-channel attacks leverage physical phenomenon not originally considered in the design of a system. The security literature has primarily studied electromagnetic radiation and power side-channels, which have been shown to be effective for breaking unsecured hardware implementations of cryptography algorithms [10]. CPMBs have unique physical characteristics that may be exploitable, such as temperature, fluorescence, and chemical residues.

2.2.2 Threat Models

Following the conventions of [11], we make a distinction between technical and operational abilities of an attacker. Technical abilities describe the knowledge an attacker has about how the microfluidic platform works and their capability to extract this information based on experimentation. Operational capabilities describe the method by which an attacker can carry out the attack. For instance, I/O ports on a microfluidic platform can be leveraged to inject malicious code, while an IP attack assumes that the attacker has access to a foundry. Note that our notions of security and trust are more general than those described in [11] since we consider IP-based attacks. We describe potential threat models for researchers to consider, organized by attacker location.

- *Manufacturing-level* threat models are a result of the untrusted supply chain. Biochip platform designers must work with and integrate components from various vendors. These vendors may be located overseas, and multiple vendors may be used simultaneously. These parties would likely be interested in carrying out IP-related attacks, and should be assumed to have considerable technical and operational capabilities since they are provided with critical design information.
- *Field-level* threats occur once the microfluidic platform is deployed and operational. Adversaries may include malicious end users who wish to modify the functionality of a device, and remote parties who are interested in the compromise of data or physical resources. These adversaries may have strong technical capabilities, especially remote parties as they may be located anywhere in the world and could be sponsored by nation-states. Their operational capabilities are more limited, as their ability to attack is dictated by the hardware and software attack surfaces available to them.

2.2.3 Motivations

Human motivations for compromising cyberphysical systems are varied and difficult to predict; however, based on the large body of evidence in related fields, we can enumerate a few common motivations that are expected to drive the previously discussed attacks.

- *Financial gain*: This motivates all the IP attacks. Evidence for counterfeiting, overbuilding, and reverse engineering abound in IC fabrication. Ransomware, identity theft, and sale of user data may drive other attacks. An interesting, more recent phenomenon is market manipulation attacks; a security research team can disclose a security flaw and cause significant changes in a company's market value. This recently occurred with the surprise disclosure of several AMD processor flaws [12], though this motivation can only be speculated.
- *Revenge*: Disgruntled employees have been behind some of the more high-profile security breaches in recent years [13]. These attackers are embedded within an organization, and once wronged, leverage their access and knowledge to devastating effect.
- *Politics*: The appearance of the Stuxnet worm in 2010 caused a reevaluation of the true magnitude of state-sponsored, politically motivated cyber threats [14].

As such, cyberphysical system designers would do well to consider strong adversaries motivated to steal trade secrets or cause physical, psychological, or financial harm.

- *Personal gain*: Researchers under pressure to publish may be tempted to fabricate data. Given the increasing effectiveness of methods used to detect spurious data, rogue researchers may be tempted to seek out more sophisticated fraud techniques.

2.3 Attack Outcomes

As a consequence of this design flow and the typical construction of a microfluidic biochip, many types of attacks are possible. We can broadly classify attacks according to their outcome as follows.

2.3.1 Reading Forgery

Reading forgery refers to the alteration or fabrication of sensor readings for the purposes of misleading the end user. After processing and sensing by the microfluidic platform, the computer that ultimately stores, processes, and transmits the sensor data can be tampered with. Computer-based attacks can be addressed using techniques such as message authentication codes and encryption [15], but cyberphysical integration leaves open the possibility of physical-based attacks.

Physical vulnerability will increasingly become a concern once microfluidics eventually makes the transition into portable devices that can easily be deployed in the field. A malicious end user can alter the sensor hardware by interrupting the connection between the sensor and the controller. In optical detection systems, a false image can be placed in front of the camera system to fool the controller. In a medical diagnostic application, this could lead to poor or deadly patient outcomes. Alternately, an attacker could modify sensor readings for the purpose of misleading error detection systems employed on the biochip.

2.3.2 Denial-of-Service

Denial-of-service (DoS) is a class of attacks that violates the availability of a system by making it unusable. DoS is more commonly experienced on a day-to-day basis as an annoyance on the internet. In a microfluidic biochip, there is potential for much greater harm. Microfluidic devices are tasked with handling reagents that may be expensive to acquire and samples that may be difficult, or even impossible, to replace. A variety of naturally occurring hardware faults such as surface contamination compounds the difficulty of ensuring reliable operation [16]. DoS is difficult to defend against since an attacker only needs to exploit a single vulnerability, and the attack can be carried out at any phase of the design flow. The forces that make DoS pervasive in computer security today—e.g., homogenization of computing platforms, and DoS-as-a-service—will also play an unwelcome role in the deployment of microfluidic biochips if care is not taken during the design phase.

2.3.3 Modification of Functionality

Modification of functionality is a threat that causes a CPMB to behave in an unintended way. There is a subtle distinction between denial-of-service and modification of functionality. In the former, the device ceases to be operational outright, while in the latter, the device continues to function but perhaps in an unanticipated or degraded way. A device that doesn't perform up to specification results in frustration for the end user, and potentially lost market share for the manufacturer. We consider abuse on the part of the end user to be a possible location for functional modification attack; if an end user wishes to bend the microfluidic platform to her will in order to achieve a desired result, this is a violation on the integrity of the assay. On a larger scale, it also violates the trust in the application in which the platform is deployed. A malicious end user presents a difficult threat, since the design of tamper-resistant devices from physically invasive attacks has been known to be elusive [17].

2.3.4 Design Theft

Design theft is a broad category of threats related to the intellectual property and information used to fabricate cyberphysical microfluidic devices. In the microfluidic domain, the intellectual property exists as the fabrication masks, materials, and processes used to build the substrate upon which fluids are manipulated, as well as the protocols, i.e., instructions for how to carry out the biochemical assay. Extensive research and development is required to produce these protocols, and companies have an interest in recouping their investment by preventing unauthorized use. Design theft has the potential to upend an industry poised for major growth in the coming years; the market for LOC technology in general reached $3.9 billion in 2014 and is projected to grow to $18.4 billion in 2020 [18]. Intellectual property disputes in microfluidics arise occasionally [19], wasting a company's financial resources that could be better spent elsewhere. Additionally, there may be little legal recourse when the perpetrators of design theft are located overseas.

The semiconductor industry suffers from supply chain problems such as overbuilding, IP theft, and counterfeiting [20], and it is reasonable to expect the same issues to plague the microfluidics industry since many fabrication techniques are derived from semiconductor processes. As the complexity of microfluidics increases, it is anticipated that the supply chain will become more horizontally integrated. This leaves open questions about the trustworthiness of the design flow [6, 21].

IP theft can even occur at the end user, since processing of fluids is often observable. A typical microfluidic device is fabricated with a transparent top plate since the final result is to be sensed by an imaging sensor. It is then trivial to observe the movement of fluids on the chip and then reverse engineer the protocol based on those observations. The problem is compounded by the low complexity and operating speed; typical DFMBs operate on the order of hundreds of hertz [22].

2.3.5 Information Leakage

Information leakage is the unauthorized dissemination of private or sensitive data. Untrusted supply chains provide numerous attack vectors for information leakage. Examples of privileged information include patient data, proprietary biochemical protocols, and secret keys. Many microfluidic biochips are intended to be deployed in medical diagnostic settings, so sensitive patient data must be correctly handled. The horizontal IC supply chain has made research in hardware Trojans a top priority; Trojans have been demonstrated to be compact while being able to disclose sensitive information such as secret keys used in cryptography [20]. Since many of the components and techniques used in constructing microfluidic devices borrow directly from the IC industry, leakage of information from compromised components is also a concern. And the unique physics of microfluidics may give rise to exploitable side-channels.

2.4 Actuation Tampering

An actuation tampering attack is a malicious modification of the signals used to control a biochip, and was first reported in the context of DMFBs with altered actuation sequences [23]. The attack describes the mechanism by which a malicious adversary can achieve outcomes such as denial-of-service or result manipulation. Actuation tampering can be carried out through many different attack vectors, including alteration of data in program memory, modification of the software used to generate actuation sequences, or physical injection of hardware faults. Flow-based biochips are also susceptible to actuation tampering through their pneumatic control valves.

The general applicability and effectiveness of actuation tampering attacks makes it attractive for malicious adversaries. The hardware security design techniques covered in this book are countermeasures against actuation tampering attacks.

To highlight the threat posed by such an attack in a CPMB, we now examine how digital polymerase chain reaction (dPCR)—a major application area for microfluidics—can be undermined.

2.4.1 Undermining Digital Polymerase Chain Reaction

Digital polymerase chain reaction (dPCR) is a relatively new method for quantifying and amplifying nucleic acids in a DNA sample [24, 25]. This method differs from traditional PCR techniques in that the sample must be split into multiple small volume reaction chambers. dPCR offers numerous strengths such as tolerance against inhibitors, lack of standard curves, and the ability to provide absolute, rather than relative, quantification [26]. In this section, we describe the concepts behind dPCR and how a commercial microfluidic platform designed for dPCR can be compromised such that the distribution of target DNA among the biochip chambers is biased. Since the dPCR reactions and valve actuations occur inside of a benchtop device, the end user would be oblivious to the attack unless the biochip was screened and analyzed for statistical anomalies—defeating the purpose of benchtop automation entirely. We then discuss the implications for copy number variation studies and research integrity in general.

2.4.2 Attacks on Commercial Microfluidic Platforms

Figure 2.1 illustrates the typical construction of a commercial chip-based dPCR microfluidic platform. A disposable chip contains the reaction chambers with inlets for samples and reagents. The chip is loaded into the platform which contains an array of sensors, actuators, and an embedded computer. The computer controls the on-chip valves to create the multiple small reaction chambers, and then cycles the temperature to carry out PCR reactions. Integrated fluorescence detectors send the result of the experiment to the embedded computer, which then either outputs the data to an integrated display or saves it to file. The microfluidic platform workflow automates many processes that were formerly conducted manually. As such, user error drops precipitously. *However, implicit in this operational protocol is trust that the devices will conduct the experiment with integrity, as the end user is not involved in any steps between the sample preparation and the final readout.*

If an attacker is able to tamper with the actuation of the microfluidic biochip, the distribution of DNA samples may be biased or the PCR reaction may be inhibited, leading to incorrect estimates of the true target DNA concentration. We studied a commercially available microfluidic platform and found that its structure closely matched that described in Fig. 2.1. The USB, serial, and Ethernet ports present an open attack surface. If these ports are unsecured, an attacker could load malicious software. We found that this particular platform used an off-the-

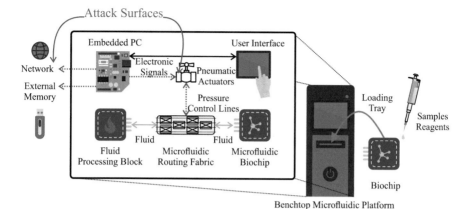

Fig. 2.1 Commercial microfluidic platforms are offered as benchtop instruments with reloadable trays. Typical construction consists of an embedded PC connected to an array of pneumatic actuators, pressure sensors, and possibly a barcode reader for automating setup and data collection. Connectivity is provided for advanced data collection capabilities, firmware updates, or reprogramming

shelf embedded single-board computer, with the custom software loaded onto a removable CompactFlash (CF) card. A platform that is physically vulnerable could be compromised simply by replacing the CF card with a malicious one. Alternately a remote party could leverage the network connectivity to assume control of the computer.

Once the controller is compromised, an attacker would be able to induce partial failure in the pneumatic actuators. The control signals could be varied to either shorten the priming time, or output a control signal with an intermediate value. The opening of elastomer valves responds linearly to pressure variations for the majority of their operating range [27] (Fig. 2.3). Therefore, the flow rate of sample into all the reaction chambers would be disrupted and the assumption of a Poisson process with fixed parameters would be violated.

2.4.3   dPCR Background

The operating principle of dPCR is based on randomly partitioning the DNA sample into multiple small reaction chambers (Fig. 2.2). These reaction chambers can be physically realized in an array (chip-based dPCR), or can be generated by encapsulating the samples in droplets generated in an oil emulsion (droplet dPCR). The PCR reaction is carried out on all of the partitions to amplify the target DNA sequence, and is then read out by a fluorescence detector. The proportion of positive

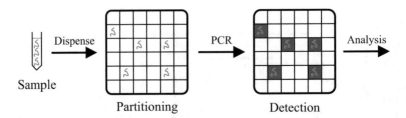

Fig. 2.2 Digital PCR works based on the random partitioning of a sample into a large number of reaction chambers. The concentration of the target DNA is estimated based on the observed positive reactions

to negative reactions can be used to calculate the number of target DNA sequences in the sample. The idea is that the random partitioning of samples will follow a Poisson distribution, and the estimated number of target DNA molecules $(\hat{M})$ can be calculated as

$$\hat{M} = -\ln(1 - \hat{H}/C) \tag{2.1}$$

where $\hat{H}$ is the observed number of positive chambers and C is the total number of chambers [28]. The measurement of the positive chambers is subject to uncertainty from sources such as inconsistent chamber volumes and non-random distribution of molecules, which has so far limited the deployment of dPCR for diagnostic applications [29].

Microfluidic technologies have lent themselves to dPCR methods, enabling applications such as studies on copy number variation and drug metabolism [28]. These devices are currently marketed for research use only, with diagnostic applications expected to occur only after the technology matures further [26, 29].

2.4.4   Simulated dPCR Attack Study

We demonstrate the results of an attack through a large-scale simulation study. A dPCR experiment can be simulated by randomly assigning M molecules into $C \times K$ reaction chambers, distributed over K number of panels. That is, for each molecule, we select one of the reaction chambers with uniform probability and assign the molecule. We then form the estimate of the true molecule concentration based on the observed number of positive reaction chambers $\hat{H}$; in this simulation, we assume the detection works perfectly. If we use the parameters provided in [28] for theory verification, we have $M = 400$, $C = 765$, and $K = 70,000$. Figure 2.4a shows a histogram of the observed $\hat{H}$ over all the K panels. To model a dPCR under a flow restriction attack, we assume that the chambers are arranged as 45 rows by 17

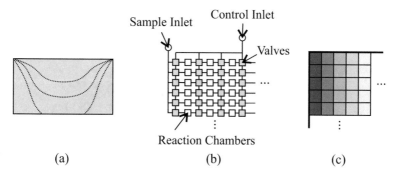

Fig. 2.3 Effect of actuation tampering on dPCR. (**a**) Cross section of a valve showing some possible deflections. Valve opening response is linear with pressure. (**b**) A DNA sample must be randomly partitioned through a grid arrangement of reaction chambers separated by valves. (**c**) Partial closure of the flow valves would increase the difficulty for fluids to flow to later stages. We represent this schematically as a grayscale gradient across the biochip. Chambers closer to the input are more likely to contain target DNA, indicated by a darker shading

columns in the same structure shown in Fig. 2.3b. We also assume that when under attack, the flow of the fluid is restricted through the columns, and some reaction sites will be more likely to contain target molecules. We model this by biasing the reaction chambers such that chambers closer to the inlet are more likely to contain sample than those farther away (Fig. 2.3c). We used a linear biased pmf $p(x) = 1/9 - (6.536 \times 10^{-3})x$, where $x \in \{0, \ldots, 16\}$ indexes the columns. Figure 2.4b shows a histogram of the results. We see that an attacker can change the mean detected number of molecules by 5.96%, with just a slight linear bias in the experiment. Thus, an attack can nudge the results to yield a false estimate.

2.4.5 Implications for Copy Number Variations

Copy number variations (CNVs) are differences in the number of structural repeats in sections of the genome [30–32]. The sensitive and accurate detection capabilities of technologies such as dPCR have enabled the study of CNVs, promising insight into the role these small variations play in genomic diseases such as autism and Crohn's disease. An attacker who compromises the microfluidic platform used to carry out dPCR would be able to influence the number of positive reactions, and thus influence the copy number ratios calculated in disease studies. Without the correct copy numbers, positive associations between these genome variations and diseases cannot be made. Worse yet, incorrect associations may be generated. Spurious associations will preclude the development treatments.

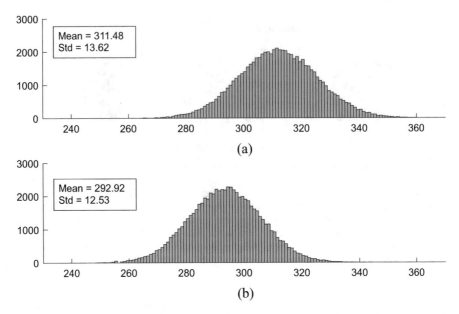

Fig. 2.4 dPCR simulation results. (**a**) Histogram of observed positive chambers $\hat{H}$ from standard dPCR experiment. (**b**) Histogram of $\hat{H}$ when valve actuation is attacked. Thus, a linear bias on the probability that molecules can flow along the biochip causes a shift in the estimate of molecules $\hat{M}$

2.4.6   Discussion

An attacker may be motivated to tamper with research equipment rather than completely fabricating data in an attempt to provide more convincing evidence that the experiments were actually carried out. Attacks on research instrumentation threatens to nullify recent efforts made to increase the quality and reproducibility of research. Specifically within dPCR, it has been noted that many researchers are not even aware of the basic methods and pitfalls of the technique—the Minimum Information for Publication of Quantitative Digital PCR Experiments (digital MIQE) guidelines were published in 2013 specifically to address these issues [33]. However, a researcher could fully comply with the standard and still release unreproducible results. Consequently, time and funding must be wasted in identifying these spurious results.

Occasionally, scientists or lab technicians are motivated to fabricate data for financial gain or bolstering their publication record. Besides blemishing the scientific literature, these violations of research integrity can have a real impact on everyday citizens (see Sect. 2.5.3). Currently, dPCR is only used within research settings due to the cost and resource requirements of the equipment involved. It is expected that dPCR microfluidics technology will mature such that it will be an attractive platform for diagnostic applications. In this case, the potential security implications of an attack on the dPCR platform could threaten the well-being of patients. An attacker could influence the decision making of a healthcare provider by skewing the diagnostic results in such a way that it is within the realm of possibility.

2.5 Challenges and Opportunities

To further motivate work in microfluidic security, we now provide comprehensive studies on application areas that either pose a significant challenge or present an opportunity for innovative research.

2.5.1 Patient Data Privacy

Medical diagnostics forms a key application area for microfluidics research. The ability to rapidly diagnose diseases and multiplex assays, especially in remote settings outside of clinics and hospitals, will be a boon for patient empowerment and improved quality of care. However, the proliferation of cheap untrusted devices will lead to a multitude of privacy violations if preventative measures are not taken. Sensitive information in a microfluidic device can include data collected after processing of the fluids and personally identifying metadata. Already, confidentiality of patient data is a concern for medical device makers [34]. Many of these same concerns and threats will be directly applicable to cyberphysical microfluidic devices.

Biochips used in the processing of biological samples would ideally undergo a certified disposal process, but regulation ensuring safe disposal is fragmented between states and the Environmental Protection Agency [35]. Additionally, security is not part of the protocol; biohazard collection bins are often located in patient rooms in plastic bins that are physically unsecured. Disposal methods designed for sterilization such as incineration, autoclaving, and microwaving are highly destructive and would likely be effective for eliminating sensitive data. However, these methods have downsides such as energy, space, and cost requirements for specialized equipment and the production of pollutants. No disposal regulations yet exist for biochips and any future legislation will likely take time to implement. And diagnostic devices deployed in remote settings cannot depend on end users to follow safe disposal practices, even if standards are published and legislated. Designers should ensure that they are able to protect sensitive information once the device enters the disposal process, even if no physical security measures are in place.

2.5.2 Defense and Public Safety

Microfluidics provides an ideal platform for defense and public safety applications. The advanced fluid handling capabilities, as well as the advantage of having a high surface-area-to-volume ratio, provide the ability to conduct a plethora of unique chemical reactions on-chip for the purposes of analyzing compounds. Examples of microfluidic devices engineered for public safety applications include monitoring of pollutants in the environment [36, 37], detection of pathogens or antibiotics for food safety assurance [38] as well as detection of chemical and biological weapons [39].

One of the most critical steps in effective defense against biological weapons is early detection of toxins and pathogens. Early detection permits treatment before the full onset of symptoms, while also protecting against those in the population who may have compromised immune systems such as the elderly. Thus it is critical that systems designed for detection be able to function continuously and reliably.

The real threat of biological warfare is difficult to assess, since 170 countries have either accessed or ratified the 1972 Biological Weapons Convention treaty [40]. It is entirely speculation as to whether these countries today are actively developing or harboring biological weapons. Still, events in Syria have shown that biological weapons are being developed and deployed: The nerve agent Sarin was released in a weaponized form, killing and injuring scores of civilians. The government of Syria has denied involvement, calling the story a fabrication [41]. Secure environmental monitoring devices enabled with microfluidics could potentially provide early warning and evidence of an attack in a time when the basic facts about real events are being debated.

Entities interested in ensuring the maximum effectiveness of their biological weapons and plausible deniability will also be interested in disabling detector systems. Nation-states are among the few entities capable of both developing large stockpiles of biological weapons and being sufficiently motivated to use them. Therefore, designers of defenses must assume that attackers possess a high level of technical sophistication.

Two methods for detecting biological weapons on-chip are immunoassays and nucleic acid testing [39]. Immunoassays determine the amount of analyte in a solution through the binding of an antibody to one specific macromolecule, while nucleic acid testing relies on PCR to amplify and identify pathogenic DNA. Several DMFB-based devices have been reported in the literature for both types of detection methods. Actuation tampering can be achieved through software means or through the payload of a hardware Trojan [42]. In continuous-flow biochips, valves are computer controlled and are thus susceptible to the same types of actuation tampering attacks. Subsequently, fluids may inadvertently mix or cause premature failure.

2.5.3 Research Integrity

The use of microfluidics in the chemical and biological sciences as a basic research tool is beneficial in terms of increasing throughput and automating complex processes, which promotes reproducibility and reliability while reducing operator errors. In fact, it has been speculated that microfluidic techniques could potentially replace the flask [43, 44]. Despite these upsides, there is a fundamental disconnect between microfluidics research within the contexts of engineering versus biology and medicine. The more prolific engineering microfluidic journals are concerned with demonstration of novel technologies and techniques, which are fabricated as proof-of-concept devices [45]. Biological and medical researchers are slow to adopt these technologies, and widespread adoption of microfluidics continues to be limited. Studies have noted that microfluidic researchers would benefit from developing novel protocols and applications rather than merely reproducing established laboratory procedures [7]. Security and trust may be one such application.

As shown in the dPCR case study in Sect. 2.4.1, microfluidic platforms could provide a sophisticated avenue for rogue researchers to commit research fraud with plausible deniability. While it is true that peer review and research reproduction can naturally screen for fraud, these processes are reactionary. This leads to cases such as that of Haruko Obokata [46, 47]—a disgraced researcher whose mentor committed suicide after her *Nature* papers [48, 49] were retracted—where research fraud incurs a human toll.

Another example of fraud took place in 2010, when it was revealed that a third-party research laboratory tasked with FDA drug screening was essentially fabricating data over a span of years [50]. This data was used to win approval for drugs, nearly 100 of which had been placed on the market. And despite these revelations, many of these drugs in question remain on the market in the interest of the drug makers involved. If the instrumentation used to carry out these tests had featured secure and trustworthy microfluidic technologies, perhaps the situation could have been avoided entirely.

Could some type of built-in integrity check used in the scientific instrumentation prevented the commitment of such high levels of fraud? The FASEB Journal makes another compelling case for preventative measures: "The more the research community responds after the fact to incidents that diminish trust, the more it leaves to chance the public's support for its work." [51].

Ambitious goals such as upholding research integrity and attestation of sensor readings are functions that cannot be realized through standard laboratory techniques; breakthroughs in security and trust validation will clearly demonstrate the superiority of lab-on-a-chip technologies. Laboratory equipment on which experiments were carried out could be monitored using cyberphysical microfluidics to provide assurances of accountability. To protect the privacy of researchers, only the information relevant to the study at hand would be logged for third parties to scrutinize. The use of microfluidics as a platform for conducting research is attractive in this regard—it allows the precise manipulation of fluids while integrating sensors for real-time tracking of experimental steps. Furthermore, advances in compressed sensing [52] could provide a means for efficiently capturing the progression of an experiment.

2.5.4 Blood Diagnostics

Theranos is a health-technology company that produced *Edison*, a microfluidic blood testing platform, in the mid-2010s. It had been touted as revolutionary for requiring only a few drops of blood while driving down costs to truly affordable levels. However, the company faced criticism and negative press over the scientific basis of its technology. The criticism proved to be well-founded in 2016, when Theranos had to recall several years' worth of blood test results [53]. Customers faced the reality of having had their medical care dictated by incorrect laboratory test results [54].

The technical details of the Theranos case are not available, but we can investigate the compromise of a widely used test for blood diagnostics: in vitro measurement of glucose [55]. Diabetic patients must undergo regular glucose tests for proper monitoring. Based on the blood glucose level, the amount of insulin to be injected into the patient is determined. By using automated microfluidic biochips, bedside glucose test can be seamlessly performed, allowing rapid, low-cost, and efficient quantification [56]. Figure 2.5 shows a graph representation, known as a sequencing graph, of the glucose-measurement assay.

Typically, this assay measures the glucose concentration level in a blood sample by constructing the glucose calibration curve (Fig. 2.5) via serial dilutions of the standard glucose solution. The X-axis represents the different concentrations formed by these dilutions (in mg/dL) and the Y-axis represents the rate of reaction quantified by the change in absorbance degree reported as AU/s (absorbance unit per second). This curve helps to interpolate the concentration of the glucose sample under test, and therefore, attaining the highest possible degree of accuracy is a must.

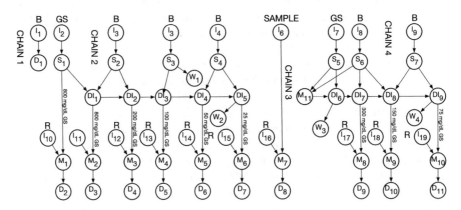

Fig. 2.5 Golden assay sequencing graph. B is the 1:4 mL buffer droplet, sample is the 0:7 mL glucose sample droplet, R is the 0:7 mL reagent droplet, GS is the 1:4 mL 800 mg/dL glucose solution droplet, and W is the waste droplet. D, DI, S, M, and I are the detection, dilution, splitting, mixing, and dispensing operations, respectively

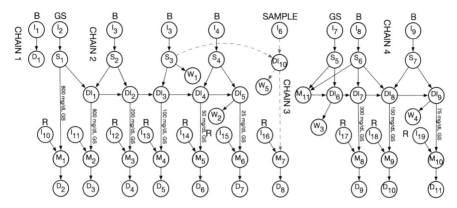

Fig. 2.6 Attack 1: Waste buffer droplet S3 dilutes the sample in Dl10, which is a direct attack on the final result. Dotted lines illustrate deviations from the golden assay

In Ali et al. [23], reported two practical result-manipulation attacks on microfluidics-based glucose tests. The first attack (Attack 1) tampers with the assay result by changing the concentration of the glucose sample as shown in Fig. 2.6. The thick dotted lines show the changes in the sequencing graph compared to the golden sequencing graph (Fig. 2.5). The waste buffer droplet W_1 generated from S_3 is mixed with the glucose sample droplet of I_6 and then diluted in Dl_{10}. Since the concentration of the glucose sample is halved, the result of the assay execution will be wrong. Note that the user is unaware that a waste buffer droplet is used for tampering with the sample concentration.

The second attack (Attack 2) is triggered by tampering with the sequencing graphs for reaction chains 2 and 4 to generate a malicious calibration curve. The two waste buffer droplets generated from D_1 and S_3 (Fig. 2.7) are used for this purpose.

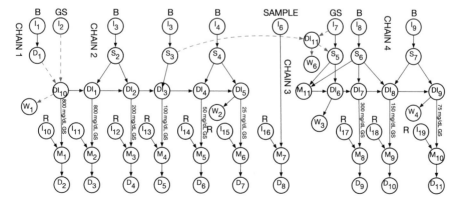

Fig. 2.7 Attack 2: Discarded buffer droplets D1 and S3 are mixed with the droplets of I2 and I7, respectively. Reaction chains 2 and 4 are subsequently diluted, resulting in an invalid calibration curve

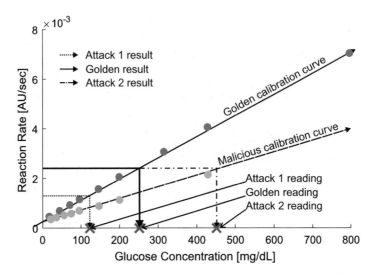

Fig. 2.8 Glucose assay results are generated against a calibration curve. Attack 1 alters the sample concentration and is interpreted against the golden calibration curve. Attack 2 alters the calibration curve, and leads to misinterpretation of the final result

The malicious sequencing graph for such an attack is shown in Fig. 2.7. The thick dotted lines show the changes with respect to the golden sequencing graph. The waste buffer droplet (after D_1) in the reaction chain 1 is merged with the glucose solution (the droplet generated from I_2) in the reaction chain 2, diluting the entire reaction chain 2. The glucose solution concentrations in the reaction chain 2 are reduced to (400, 200, 100, 50, 25, 12.5 mg/dL) half of their golden values. A similar effect can also be seen in reaction chain 4, where the waste buffer droplet generated from S_3 is mixed with the glucose solution droplet generated from I_7. The dotted curve in Fig. 2.8 shows the malicious calibration curve generated by Attack 2.

2.5.5 Drug Doping

Numerous high-profile cases of performance-enhancing drug use have appeared over the years, resulting in the relinquishment of titles and suspension from competition. Many of these accusations of drug doping came after failing a test, and later, having admitted to accepting performance-enhancing drugs. And in some cases, it has been revealed that doping was being conducted on a large-scale systemic basis such as in East Germany and more recently, Russia [57, 58]. However there is an open question as to how effective these tests really are. What is the false positive rate? What is the false negative rate? Can the agencies and instruments used in drug screening be compromised? And are there problems with the statistical reasoning used in these drug tests [59]?

The World Anti-Doping Agency (WADA) was founded in 1999 in order to address anti-doping policies and regulations, with functions including monitoring and education of athletes. WADA publishes the International Standard for Testing and Investigations (ISTI), which describes, among other things, procedures on athlete notification for conducting out-of-competition testing. Athletes have voiced their ire about the procedures [60]. POC diagnostic devices could be used to enable more convenient, less invasive testing methods, but such a usage model leaves the possibility of athletes and trainers tampering with the devices or even using samples not belonging to the athlete-under-test. Furthermore, Section 10.1 of the ISTI states that "Samples collected from an Athlete are owned by the Testing Authority for the Sample Collection Session in question." This potentially poses an ownership and privacy issue, as the drug screening process is supposed to only answer the question of whether any banned substances were used by athletes. Relinquishing whole samples could potentially leave open the possibility of unrelated information being determined and released. A trustworthy POC testing platform would be beneficial for both the integrity of sport and the convenience, comfort, and privacy of athletes.

Modification of functionality can be achieved through actuation tampering, production of counterfeit devices, or even negligence on the part of the end user. Hardware and software components are both vulnerable to modification, and can occur along any stage of the development cycle. A nation-state that engages in systemic doping fraud would conceivably leverage their technical capabilities to modify the functionality of diagnostic devices. Security and trust techniques targeted for such applications will necessarily require strong threat models.

2.5.6 DNA Forensics

DNA forensics is the application of DNA fingerprinting techniques for identifying individuals in criminal investigations. In a traditional DNA forensic flow, a DNA sample is first collected, then extracted, amplified, detected, and either stored or destroyed [61]. Each of the DNA processing steps must be performed at a specially equipped forensics laboratory. In the USA, regulation of forensics laboratories is fragmented, with only a handful of states requiring accreditation for DNA forensics [62]. Even with accreditation and government oversight, issues with the DNA forensics process have been identified such as contamination of the samples and negligence by analysts [63–65]. Development and adaption of microfluidic technology for DNA forensics has been motivated by its potential to address the shortcomings inherent in the traditional laboratory flow, as well as to decrease turnaround time and cost.

Microfluidic devices implementing processes such as cell lysis [66, 67], DNA purification [68], extraction [69], polymerase chain reaction (PCR) [70], and detection [71, 72] have been successfully demonstrated. However, the current state-of-the-art does not permit the integration of all these steps in a single, portable lab-on-a-chip device [73]. Commercial deployment of these technologies is limited at the time of this writing [7]. It is expected that microfluidic technology will eventually mature to the point where a device could be deployed at a crime scene, which is appealing from both a reliability and security standpoint; automation and the reduction in the number of steps involved in the analysis decreases the likelihood of contamination or intentional fouling [61]. On the other hand, the new technology may present new unanticipated attack vectors. The nascency of microfluidics in DNA forensics presents an opportunity to incorporate security and trust in critical applications such as criminal investigations and intelligence gathering.

2.6 Summary and Conclusion

Security and trust of CPMBs involves new physical modalities and exciting new application areas while consisting of elements from traditional hardware security, embedded cyberphysical system security, and more traditional information security practices. The problems are challenging due to the complex interdisciplinary nature of the systems involved, but these same novel attributes give rise to unique countermeasures. Furthermore, we conclude that there exists a need for *hardware* security: the study of vulnerabilities that arise as a consequence of hardware design decisions, and the development of countermeasures in both hardware and software.

We close by summarizing this chapter in several tables. Table 2.1 discusses some potential security issues of selected microfluidic platforms that have been published, commercialized, or made available under open-source licenses. This table is far from comprehensive, and interested readers are recommended to read recent review papers for more information on how state-of-the-art microfluidic biochips work [7, 78]. Table 2.2 shows some example scenarios under different attack surfaces and attack outcomes. Table 2.3 shows various unique properties of microfluidic biochips and how they can be leveraged or abused. Security and trust issues organized by location in the biochip design flow are shown in Table 2.4.

Table 2.1 Comparison of selected cyberphysical microfluidic platforms and security implications

Company	Platform	Description	Applications	Security implications
Fluidigm	BioMark HD	Samples and reagents loaded onto proprietary integrated fluidic circuit (IFC) cartridges, which are placed into the platform for mixing, thermal cycling, and fluorescence CCD detection	Single-cell genomics, real-time PCR, library prep	Requires standard PC connection for analysis, which is a potent attack vector for malware. Reliance on standard electronic components such as CCD cameras and microcontrollers leaves open the possibility of physical tampering and hardware Trojans
Baebies	SEEKER	High-throughput benchtop analyzer measures activity of lysosomal enzymes in dried blood spot specimens using digital microfluidics. Spot Logic Software running on connected PC controls platform and displays results	Newborn screening for lysosomal storage disorders	Requires standard PC connection for analysis, which is a potent attack vector for malware. Proprietary electrochemical detection mechanisms increase difficulty of physical reading forgery attacks
Illumina	NeoPrep	Benchtop EWOD digital microfluidics-based library prep system. Acquired from Advanced Liquid Logic, currently discontinued	Next-generation sequencing library preparation for scientific research	Runs Windows Embedded Standard 7, which may pose a security risk. Reprogrammable nature of digital microfluidics can be abused to modify functionality
GenMark	ePlex	Single-use cartridge "panels" contain nucleic acid extraction, amplification, detection, barcodes, reagents. Panels loaded into ePlex benchtop system for automated processing and data collection with proprietary eSensor electrochemical detectors	Infectious disease testing	Data interfaces to servers through laboratory information system (LIS) interface, which is not standardized in industry and may not implement modern security measures [74]. Cartridges can be counterfeited and barcodes can be spoofed

(continued)

Table 2.1 (continued)

Company	Platform	Description	Applications	Security implications
SensoDx	SensoDx	Samples loaded onto blister pack cartridges which contain reagents. Blister pack loaded into platform which actuates fluid flow, detects the result, and displays data. Embedded PC leverages machine learning techniques to interpret data	PoC diagnostics	Platform is constructed using disparate off-the-shelf components and custom parts from overseas fabricators. Designs can easily be stolen, overproduced, and sold for less. Machine learning techniques may be susceptible to poisoning attacks [75]
Abbott Laboratories	i-STAT	Handheld blood analyzer. One of the earliest microfluidic point-of-care diagnostic devices. Cartridge with blood sample is processed through capillary-driven microfluidics and detected with thin-film electrodes	PoC blood diagnostics	Data can be loaded into i-STAT Downloader tool, or wirelessly transferred to PC for management in Abbott Info HQ Manager; marketing materials do not indicate any security measures such as encryption to protect patient data
Auryn.bio	OpenDrop	Open-source digital microfluidics laboratory. DC actuation. PCB-based electrodes with PC controller software [76]	Democratization of biological science	Open-source software potentially allows independent auditing for security vulnerabilities, but some may escape detection (see: Heartbleed)
Sci-Bots Inc.	DropBot	Open-source digital microfluidics. Electrical impedance based droplet sensing. AC actuation. Requires laboratory high-voltage amplifier. Spun off work at Wheeler Laboratory [77]	Chemistry and life sciences	See above

Table 2.2 Example attack scenarios

Attack surface	Ex.	Reading forgery	Information leakage	Denial-of-service	Modification of functionality
Indirect physical access	USB, Ethernet, Flash port, cartridge loading tray	Deliver malware to modify actuation sequences	Read files off network due to unsecured ports or poor access control policies	Load modified actuation sequences through Flash port	Mislabeled or forged barcode data on cartridge
Direct physical access	Clock/power glitching, tamper with control signals	Inject false sensor readings through the sensor signal lines	Probe on-board communication channels, observe droplet movement	Contaminate electrodes, samples, or reagents	Overheat biochip to degrade reliability
Wireless access	Wi-Fi, Bluetooth	Modify packets en route to central server through rogue access point	Wi-Fi packet sniffing	Radio jamming	Deliver malware through administrative interface
Standard interfaces	Touchscreen, keyboard	N/A	Unauthorized disclosure of information through privilege escalation	Input operating parameters that cause damage to platform	N/A
Side-channel	Electromagnetic, power, chemical, temperature	N/A	Analyze chemical residue on discarded biochip	N/A	N/A

Table 2.3 Cyberphysical microfluidic properties and corresponding security opportunities

Property	Description	Security challenges and opportunities
Slow operation	Due to physical limitations, DMFB actuation sequences are necessarily applied at relatively low frequencies, on the order of kHz or less [79]. Chemical time constants are orders of magnitude slower than in electrical domain	Slow devices make monitoring easy, and can limit the scope of attacks that drastically alter execution times. Many stealthy attacks are still possible [23], but defenses can take advantage of low speed to make executions more tamper-evident
Visible droplet movement	Execution of assays on biochips is often visible with the naked eye	Can enable technologies to monitor assay progression such as randomized checkpointing with CCD cameras [80, 81]
Reconfigurability	Routing crossbars [82] and general-purpose biochips allow the same hardware to be used for different protocols	Arbitrary droplet movement allows attackers to modify assay to cause denial-of-service or subtle errors [23]. Routing crossbars are more constrained but could still have unintended operation. Reconfigurability and security trade off [83]
Sensor feedback	A variety of sensors can be integrated onto a biochip to provide feedback on the state of the chip or to measure some property of a droplet, such as for colorimetric assays	Depending on how sensor is integrated, the reading may be susceptible to tampering. Sensor readings have no binding with patient metadata. Sensor physical unclonable function is a promising security primitive for "encrypting" readings while ensuring authenticity [84]
Relatively large scale	Loading of reagents/samples onto biochips must interface with the "macro" world, meaning that devices are often much larger than micrometer scale [7]	Interfaces designed to be serviceable by humans can be physically tampered with. Devices need to be designed with tamper resistance in mind, which may occur in a macro/mechanical domain [85], or using sensor physical unclonable functions [84]
Low-cost fabrication	Microfluidics are increasingly becoming accessible, with new low-cost techniques being developed that utilize PCBs and off-the-shelf components	Lowering the barrier for entry leads to lower reliability, quality, or incorrect measurements. PUFs can be used to authenticate genuine devices [86, 87]
High-voltage actuation	Actuation voltages used in electrowetting-on-dielectric devices are often large, in the range of hundreds of volts	High voltages necessarily produce more electromagnetic noise. This can cause leakage of information in the form of a side-channel; may be difficult to control until operating voltages can be scaled down

Open-source platforms	DropBot [77], OpenDrop [76], UCR Static Synthesis [88, 89] are among the first platforms freely available under open-source licenses	Open-source hardware and software provides the ability to independently audit the design to see if it follows best practices or contains Trojans. Adoption appears to be limited at present. Platforms are primitive as compared to the state-of-the-art (e.g., no error recovery)
Automation	Biochips have a rich, established literature on design automation techniques	A platform that is automated implies the presence of some computational ability; some portion of this computing power can be redirected for security purposes
High-level synthesis	Assay specifications must go through a synthesis toolchain before usable actuation sequences can be applied to DMFB	The many software layers in the synthesis toolchain, down to hardware actuation sequences present opportunities for tampering. Encryption and obfuscation are possible candidates to protect assays from tampering, or from intellectual property theft [6, 90]
One-time use	Biochips are often designed to be single use to prevent contamination	Single-use biochips allow usage of weak PUFs, which are limited in the number of challenge-responses [86]. SensorPUFs become more feasible in the single-use model [84]

Table 2.4 Summary of security threats by location

Location	Denial-of-service	Reading forgery	Information leakage	Modification of functionality	Design theft
CAD tool vendor	Untrusted CAD tools can insert malicious behavior into the synthesized design [91]	Alternate actuation sequences can be synthesized that perform to incorrect calibration or dilutions, leading to misleading results [23]	N/A	Synthesized control sequences can be generated which perform the basic intended functions, but perhaps with degraded performance through excessive actuations	CAD tools must be given an assay protocol as input. Untrusted software could potentially steal these protocols
Biochip vendor	Hardware Trojans can be inserted at fabrication time which can cause an assay to fail due to incorrect droplet routing [20, 23]	Untrusted sensors could be integrated into the final biochip design, giving misleading results	Hardware Trojans can leak sensitive information such as patient diagnostic readings. Trigger condition can be barcode of a specific patient	Hardware layer almost always exists between the CPU and biochip hardware, or may be integrated such as in MEDA biochips [92], leaving possibility of Trojan insertion	As biochip processing becomes more advanced, horizontal supply chains become more attractive, leading to the possibility of hardware design theft [6]
Tester	Testers could not follow the test protocol, or not even test biochips before releasing for sale. High failure rates cannot definitively be tied to poor test due to inherent microfluidic failure modes	N/A	N/A	See Tester DoS attacks	Testers could steal design information if it needs to be provided to conduct a comprehensive test (e.g., from golden models)

End user	Malicious end users can cause actuation tampering attacks through fault injection [93]: power/clock glitching or manipulation of valves	Point-of-care technologies may be vulnerable to invasive physical attacks, such as sensor spoofing	Sensitive data may be trivial to collect if the biochip is visually exposed; cameras can record assays as they execute	Actuation tampering attacks can misdirect fluids or give incorrect dilutions. Negligence by poorly trained personnel can cause unintended behavior	Devices and protocols can be reverse engineered by studying the construction and related documentation/publications
Recycling and disposal	N/A	N/A	Sensitive information may exist on discarded biochips in the form of residues. Data remanence attacks may be possible on the controllers after they have been decommissioned	Discarded biochips can be harvested for resale into the gray market, leading to degraded performance	N/A
Remote party	Control software can be compromised, such that it produces incorrect control sequences. Alternately, pre-synthesized control sequences can be tampered with	Compromised controllers can cause the assay to fail and give a misleading result, force the sensor to report incorrect information, or modify the readings after they have been correctly reported	Data collected by the platform may be readable if not properly secured through techniques such as encryption	Similar to denial-of-service attacks, compromised control software can be directed to produce more subtle failures and downgrades in performance	Actuation sequences residing in memory could be reverse engineered to recover the protocol specification

References

1. The New York Times, *'You Are the Product': Targeted by Cambridge Analytica on Facebook* (The New York Times, New York, 2018)
2. The New York Times, *2.5 Million More People Potentially Exposed in Equifax Breach* (The New York Times, New York, 2017)
3. G. Zhang, C. Yan, X. Ji, T. Zhang, T. Zhang, W. Xu, DolphinAttack: inaudible voice commands, in *Proceedings of the 2017 ACM SIGSAC Conference on Computer and Communications Security* (ACM, New York, 2017), pp. 103–117
4. S. Bhunia, M.S. Hsiao, M. Banga, S. Narasimhan, Hardware Trojan attacks: threat analysis and countermeasures. Proc. IEEE **102**, 1229–1247 (2014)
5. SEMI, *IP Challenges for the Semiconductor Equipment and Materials Industry* (2012)
6. S.S. Ali, M. Ibrahim, J. Rajendran, O. Sinanoglu, K. Chakrabarty, Supply-chain security of digital microfluidic biochips. Computer **49**(8), 36–43 (2016)
7. L.R. Volpatti, A.K. Yetisen, Commercialization of microfluidic devices. Trends Biotechnol. **32**(7), 347–350 (2014)
8. F. Su, K. Chakrabarty, High-level synthesis of digital microfluidic biochips. ACM J. Emerg. Technol. Comput. Syst. **3**(4), p. 1 (2008)
9. T.M. Squires, S.R. Quake, Microfluidics: fluid physics at the nanoliter scale. Rev. Mod. Phys. **77**(3), 977 (2005)
10. D. Agrawal, B. Archambeault, J.R. Rao, P. Rohatgi, The EM side-channel(s), in *International Workshop on Cryptographic Hardware and Embedded Systems* (Springer, Berlin, 2002), pp. 29–45
11. S. Checkoway, D. McCoy, B. Kantor, D. Anderson, H. Shacham, S. Savage, K. Koscher, A. Czeskis, F. Roesner, T. Kohno, et al., Comprehensive experimental analyses of automotive attack surfaces, in *Proceedings of USENIX Security Symposium*, San Francisco (2011), pp. 77–92
12. B. Schneier, *Israeli Security Company Attacks AMD by Publishing Zero-Day Exploits* (2018)
13. A. Cardenas, S. Amin, B. Sinopoli, A. Giani, A. Perrig, S. Sastry, Challenges for securing cyber physical systems, in *Proceedings of the Workshop on Future Directions in Cyber-physical Systems Security* (2009), p. 5
14. R. Langner, Stuxnet: dissecting a cyberwarfare weapon. IEEE Secur. Priv. **9**(3), 49–51 (2011)
15. N. Ferguson, B. Schneier, T. Kohno, *Cryptography Engineering: Design Principles and Practical Applications* (Wiley, Hoboken, 2011)
16. T. Xu, K. Chakrabarty, Functional testing of digital microfluidic biochips, in *2007 IEEE International Test Conference* (IEEE, Piscataway, 2007), pp. 1–10
17. R. Anderson, M. Kuhn, Tamper resistance—a cautionary note, in *Proceedings of the Second USENIX Workshop on Electronic Commerce*, vol. 2 (1996), pp. 1–11
18. J. Evans, *Global Biochip Markets: Microarrays and Lab-on-a-Chip*, Tech. Rep. BIO049F (BCC Research, Wellesley, 2016)
19. A. McWilliams, *Microfluidics: Technologies and Global Markets*, Tech. Rep. (BCC Research, Wellesley, 2013)
20. M. Rostami, F. Koushanfar, R. Karri, A primer on hardware security: models, methods, and metrics. Proc. IEEE **102**(8), pp. 1283–1295 (2014)
21. H. Chen, S. Potluri, F. Koushanfar, BioChipWork: reverse engineering of microfluidic biochips, in *2017 IEEE International Conference on Computer Design (ICCD)* (IEEE, Piscataway, 2017), pp. 9–16
22. M. Pollack, A. Shenderov, R. Fair, Electrowetting-based actuation of droplets for integrated microfluidics. Lab. Chip **2**(2), pp. 96–101 (2002)
23. S.S. Ali, M. Ibrahim, O. Sinanoglu, K. Chakrabarty, R. Karri, Security assessment of cyberphysical digital microfluidic biochips. IEEE/ACM Trans. Comput. Biol. Bioinform. **13**(3), 445–458 (2016)
24. B. Vogelstein, K.W. Kinzler, Digital PCR. Proc. Natl. Acad. Sci. **96**(16), 9236–9241 (1999)

25. P. Sykes, S. Neoh, M. Brisco, E. Hughes, J. Condon, A. Morley, Quantitation of targets for PCR by use of limiting dilution. Biotechniques **13**(3), 444–449 (1992)
26. M. Baker, Digital PCR hits its stride. Nat. Methods **9**(6), 541–544 (2012)
27. M.A. Unger, H.-P. Chou, T. Thorsen, A. Scherer, S.R. Quake, Monolithic microfabricated valves and pumps by multilayer soft lithography. Science **288**(5463), 113–116 (2000)
28. S. Dube, J. Qin, R. Ramakrishnan, Mathematical analysis of copy number variation in a DNA sample using digital PCR on a nanofluidic device. PLoS One **3**(8), e2876 (2008)
29. J.F. Huggett, S. Cowen, C.A. Foy, Considerations for digital PCR as an accurate molecular diagnostic tool. Clin. Chem. **61**(1), pp. 79–88 (2015)
30. M. Zarrei, J.R. MacDonald, D. Merico, S.W. Scherer, A copy number variation map of the human genome. Nat. Rev. Genet. **16**(3), 172 (2015)
31. A.J. Lafrate, L. Feuk, M.N. Rivera, M.L. Listewnik, P.K. Donahoe, Y. Qi, S.W. Scherer, C. Lee, Detection of large-scale variation in the human genome. Nat. Genet. **36**(9), 949 (2004)
32. J. Sebat, B. Lakshmi, J. Troge, J. Alexander, J. Young, P. Lundin, S. Månér, H. Massa, M. Walker, M. Chi, N. Navin, R. Lucito, J. Healy, J. Hicks, K. Ye, A. Reiner, T.C. Gilliam, B. Trask, N. Patterson, A. Zetterberg, M. Wigler, Large-scale copy number polymorphism in the human genome. Science **305**(5683), pp. 525–528 (2004)
33. J.F. Huggett, C.A. Foy, V. Benes, K. Emslie, J.A. Garson, R. Haynes, J. Hellemans, M. Kubista, R.D. Mueller, T. Nolan, et al., The digital MIQE guidelines: minimum information for publication of quantitative digital PCR experiments. Clin. Chem. **59**(6), 892–902 (2013)
34. M. Zhang, A. Raghunathan, N.K. Jha, Trustworthiness of medical devices and body area networks. Proc. IEEE **102**(8), 1174–1188 (2014)
35. W.C. Blackman Jr., *Basic Hazardous Waste Management* (CRC Press, Boca Raton, 2016)
36. G. Chen, Y. Lin, J. Wang, Monitoring environmental pollutants by microchip capillary electrophoresis with electrochemical detection. Talanta **68**(3), 497–503 (2006)
37. J.C. Jokerst, J.M. Emory, C.S. Henry, Advances in microfluidics for environmental analysis. Analyst **137**(1), 24–34 (2012)
38. S. Neethirajan, I. Kobayashi, M. Nakajima, D. Wu, S. Nandagopal, F. Lin, Microfluidics for food, agriculture and biosystems industries. Lab. Chip **11**(9), 1574–1586 (2011)
39. J. Wang, Microchip devices for detecting terrorist weapons. Anal. Chim. Acta **507**(1), 3–10 (2004)
40. L.G.W. Christopher, L.T.J. Cieslak, J.A. Pavlin, E.M. Eitzen, Biological warfare: a historical perspective. JAMA **278**(5), 412–417 (1997)
41. The New York Times, *Banned Nerve Agent Sarin Used in Syria Chemical Attack, Turkey Says* (The New York Times, New York, 2017)
42. R. Karri, J. Rajendran, K. Rosenfeld, M. Tehranipoor, Trustworthy hardware: identifying and classifying hardware Trojans. Computer **43**, 39–46 (2010)
43. K.S. Elvira, X.C.I Solvas, R.C. Wootton, A.J. deMello, The past, present and potential for microfluidic reactor technology in chemical synthesis. Nat. Chem. **5**(11), 905–915 (2013)
44. J. Knight, Microfluidics: honey, I shrunk the lab. Nature **418**(6897), 474–475 (2002)
45. H.H. Caicedo, S.T. Brady, Microfluidics: the challenge is to bridge the gap instead of looking for a 'killer app'. Trends Biotechnol. **34**, 1–3 (2016)
46. D. Goodyear, The stress test, *The New Yorker* (2016)
47. Z. Schlanger, Haruko Obokata, who claimed stem cell breakthrough, found guilty of scientific misconduct, *Newsweek* (2014)
48. H. Obokata, Y. Sasai, H. Niwa, M. Kadota, M. Andrabi, N. Takata, M. Tokoro, Y. Terashita, S. Yonemura, C.A. Vacanti, T. Wakayama, Bidirectional developmental potential in reprogrammed cells with acquired pluripotency. Nature **505**(7485), 676–680 (2014)
49. H. Obokata, T. Wakayama, Y. Sasai, K. Kojima, M.P. Vacanti, H. Niwa, M. Yamato, C.A. Vacanti, Stimulus-triggered fate conversion of somatic cells into pluripotency. Nature **505**(7485), 641–647 (2014)
50. R. Garver, C. Seife, *FDA Let Drugs Approved on Fraudulent Research Stay on the Market* (2013)

51. M. Yarborough, Taking steps to increase the trustworthiness of scientific research. FASEB J. **28**(9), 3841–3846 (2014)
52. D.L. Donoho, Compressed sensing. IEEE Trans. Inf. Theory **52**(4), 1289–1306 (2006)
53. R. Abelson, A. Pollack, *Walgreens Cuts Ties to Blood-Testing Company Theranos* (2016)
54. The Wall Street Journal, *Theranos Results Could Throw Off Medical Decisions, Study Finds* (The Wall Street Journal, New York, 2016)
55. P. Trinder, Determination of glucose in blood using glucose oxidase with an alternative oxygen acceptor. Ann. Clin. Biochem. **6**(1), 24–27 (1969)
56. V. Srinivasan, V.K. Pamula, R.B. Fair, Droplet-based microfluidic lab-on-a-chip for glucose detection. Anal. Chim. Acta **507**(1), 145–150 (2004)
57. S. Ungerleider, *Faust's Gold: Inside the East German Doping Machine* (Thomas Dunne Books, New York, 2001)
58. The New York Times Company, *The Russian Doping Scandal* (The New York Times Company, New York, 2016)
59. D.A. Berry, The science of doping. Nature **454**, pp. 692–693 (2008)
60. BBC, *Athletes Air Issues Over Testing* (2009)
61. B. Bruijns, A. van Asten, R. Tiggelaar, H. Gardeniers, Microfluidic devices for forensic DNA analysis: a review. Biosensors **6**(3), 41 (2016)
62. N.R. Council et al., *Strengthening Forensic Science in the United States: A Path Forward* (National Academies Press, Washington, 2009)
63. H. Edwards, C. Gotsonis, Strengthening forensic science in the United States: a path forward, in *Statement Before the United State Senate Committee on the Judiciary* (2009)
64. J.W. Bond, C. Hammond, The value of DNA material recovered from crime scenes. J. Forensic Sci. **53**(4), 797–801 (2008)
65. W.C. Thompson, Subjective interpretation, laboratory error and the value of forensic DNA evidence: three case studies. Genetica **96**(1–2), 153–168 (1995)
66. J.M. Bienvenue, N. Duncalf, D. Marchiarullo, J.P. Ferrance, J.P. Landers, Microchip-based cell lysis and DNA extraction from sperm cells for application to forensic analysis. J. Forensic Sci. **51**(2), 266–273 (2006)
67. C.-Y. Lee, G.-B. Lee, J.-L. Lin, F.-C. Huang, C.-S. Liao, Integrated microfluidic systems for cell lysis, mixing/pumping and DNA amplification. J. Micromech. Microeng. **15**(6), 1215 (2005)
68. J.M. Bienvenue, L.A. Legendre, J.P. Ferrance, J.P. Landers, An integrated microfluidic device for DNA purification and PCR amplification of STR fragments. Forensic Sci. Int. Genet. **4**(3), 178–186 (2010)
69. L.A. Legendre, J.M. Bienvenue, M.G. Roper, J.P. Ferrance, J.P. Landers, A simple, valveless microfluidic sample preparation device for extraction and amplification of DNA from nanoliter-volume samples. Anal. Chem. **78**(5), 1444–1451 (2006)
70. J. Khandurina, T.E. McKnight, S.C. Jacobson, L.C. Waters, R.S. Foote, J.M. Ramsey, Integrated system for rapid PCR-based DNA analysis in microfluidic devices. Anal. Chem. **72**(13), 2995–3000 (2000)
71. E.T. Lagally, P.C. Simpson, R.A. Mathies, Monolithic integrated microfluidic DNA amplification and capillary electrophoresis analysis system. Sens. Actuators B **63**(3), 138–146 (2000)
72. B.S. Ferguson, S.F. Buchsbaum, J.S. Swensen, K. Hsieh, X. Lou, H.T. Soh, Integrated microfluidic electrochemical DNA sensor. Anal. Chem. **81**(15), 6503–6508 (2009)
73. R.B. Fair, Digital microfluidics: is a true lab-on-a-chip possible? Microfluid. Nanofluid. **3**(3), 245–281 (2007)
74. J.L. Sepulveda, D.S. Young, The ideal laboratory information system. Arch. Pathol. Lab. Med. **137**(8), 1129–1140 (2013)
75. M. Mozaffari-Kermani, S. Sur-Kolay, A. Raghunathan, N.K. Jha, Systematic poisoning attacks on and defenses for machine learning in healthcare. IEEE J. Biomed. Health. Inf. **19**(6), 1893–1905 (2015)
76. M. Alistar, U. Gaudenz, OpenDrop: an integrated do-it-yourself platform for personal use of biochips. Bioengineering **4**(2), 45 (2017)

77. R. Fobel, C. Fobel, A.R. Wheeler, DropBot: an open-source digital microfluidic control system with precise control of electrostatic driving force and instantaneous drop velocity measurement. Appl. Phys. Lett. **102**(19), 193513 (2013)
78. C.D. Chin, V. Linder, S.K. Sia, Commercialization of microfluidic point-of-care diagnostic devices. Lab. Chip **12**, 2118–2134 (2012)
79. V. Srinivasan, V.K. Pamula, R.B. Fair, An integrated digital microfluidic lab-on-a-chip for clinical diagnostics on human physiological fluids. Lab. Chip **4**(4), 310–315 (2004)
80. J. Tang, M. Ibrahim, K. Chakrabarty, R. Karri, Secure randomized checkpointing for digital microfluidic biochips. IEEE Trans. Comput. Aided Design Integr. Circuits Syst. **37**, 1119–1132 (2018)
81. J. Tang, M. Ibrahim, K. Chakrabarty, R. Karri, Securing digital microfluidic biochips by randomizing checkpoints, in *Proceedings of IEEE International Test Conference (ITC)* (IEEE, Piscataway, 2016), pp. 1–8
82. R. Silva, S. Bhatia, D. Densmore, A reconfigurable continuous-flow fluidic routing fabric using a modular, scalable primitive. Lab. Chip **16**(14), 2730–2741 (2016)
83. J. Tang, M. Ibrahim, K. Chakrabarty, R. Karri, Security trade-offs in microfluidic routing fabrics, in *2017 IEEE International Conference on Computer Design (ICCD)* (IEEE, Piscataway, 2017), pp. 25–32
84. K. Rosenfeld, E. Gavas, R. Karri, Sensor physical unclonable functions, in *2010 IEEE International Symposium on Hardware-Oriented Security and Trust (HOST)* (IEEE, Piscataway, 2010), pp. 112–117
85. D. Shahrjerdi, J. Rajendran, S. Garg, F. Koushanfar, R. Karri, Shielding and securing integrated circuits with sensors, in *Proceedings of the 2014 IEEE/ACM International Conference on Computer-Aided Design* (IEEE Press, Piscataway, 2014), pp. 170–174
86. C. Herder, M.-D. Yu, F. Koushanfar, S. Devadas, Physical unclonable functions and applications: a tutorial. Proc. IEEE **102**(8), 1126–1141 (2014)
87. L. Wei, C. Song, Y. Liu, J. Zhang, F. Yuan, Q. Xu, BoardPUF: physical unclonable functions for printed circuit board authentication, in *Proceedings of the IEEE/ACM International Conference on Computer-Aided Design* (IEEE, Piscataway, 2015), pp. 152–158
88. D. Grissom, K. O'Neal, B. Preciado, H. Patel, R. Doherty, N. Liao, P. Brisk, A digital microfluidic biochip synthesis framework, in *2012 IEEE/IFIP 20th International Conference on VLSI and System-on-Chip (VLSI-SoC)* (IEEE, Piscataway, 2012), pp. 177–182
89. D. Grissom, C. Curtis, S. Windh, C. Phung, N. Kumar, Z. Zimmerman, O. Kenneth, J. McDaniel, N. Liao, P. Brisk, An open-source compiler and PCB synthesis tool for digital microfluidic biochips. Integr. VLSI J. **51**, 169–193 (2015)
90. S.S. Ali, M. Ibrahim, O. Sinanoglu, K. Chakrabarty, R. Karri, Microfluidic encryption of on-chip biochemical assays, in *2016 IEEE Biomedical Circuits and Systems Conference (BioCAS)* (IEEE, Piscataway, 2016), pp. 152–155
91. M. Potkonjak, Synthesis of trustable ICs using untrusted CAD tools, in *Proceedings of IEEE/ACM Design Automation Conference* (IEEE, Piscataway, 2010), pp. 633–634
92. G. Wang, D. Teng, Y.-T. Lai, Y.-W. Lu, Y. Ho, C.-Y. Lee, Field-programmable lab-on-a-chip based on microelectrode dot array architecture. IET Nanobiotechnol. **8**(3), 163–171 (2013)
93. H. Bar-El, H. Choukri, D. Naccache, M. Tunstall, C. Whelan, The sorcerer's apprentice guide to fault attacks. Proc. IEEE **94**(2), 370–382 (2006)

Chapter 3
Prevention: Tamper-Resistant Pin-Constrained Digital Microfluidic Biochips

3.1 Introduction

Practical DMFBs often employ a technique called *pin mapping* to reduce control pin count while simultaneously reducing the degrees of freedom available for droplet manipulation [1]. Attempts to control specific electrodes as part of an actuation tampering attack cannot be made without inadvertently actuating other electrodes and causing inadvertent droplet movements, which makes the tampering evident. Thus, pin mapping can be considered as a mechanism for preventing an actuation tampering attack on DMFB.

Technologies for preventing an actuation tampering attack on a CPMB can either be direct or indirect. Direct prevention presents a mechanism or interface that makes it difficult for an attacker to achieve their desired goal. Such difficulty can be quantified through security metrics such as computation time or the technological sophistication required to break the mechanism. Indirect prevention presents a cost to the attacker. For instance, the ability to successfully attribute an attack to a particular nation, corporation, or individual would deter attackers unwilling to risk the consequences of prosecution. Leveraging pin mapping as a tamper resistance mechanism falls under the category of direct prevention.

This chapter explores the tamper resistance property of pin mapping in detail. We derive relevant security metrics, evaluate the tamper resistance of several existing pin mapping algorithms, and propose a new security-aware pin mapper with superior tamper resistance as compared to prior work. We then take steps to boost tamper resistance through the introduction of indicator droplets, as the sparsity of a given bioassay sets a fundamental limit on achievable tamper resistance.

© Springer Nature Switzerland AG 2020
J. Tang et al., *Secure and Trustworthy Cyberphysical Microfluidic Biochips*,
https://doi.org/10.1007/978-3-030-18163-5_3

3.2   Broadcast Addressing

Many post-synthesis pin mappers are based upon broadcast electrode-addressing schemes, which hardwires electrodes into pins receiving the same sequence of control signals [1]. Broadcast addressing relies on the concepts of *don't-care* values and *compatible sequences*.

Don't-cares On a DMFB grid, the movement of a single droplet requires a single pin activation (represented in the actuation sequence as a 1) to change the contact angle and initiate movement, and the deactivation (represented as 0) of the surrounding pins to ensure that the droplet does not inadvertently split. Any other electrode not directly involved in this transfer is a don't-care (represented by x), and can be held either high or low. The convention is chosen by the biochip designer, though typically it is held low.

Compatible Sequences Two electrode actuation sequences are compatible with each other if each value is either identical or at least one electrode contains a don't-care. Two compatible sequences can be combined into one by replacing don't-cares with the other electrode's actuation value. This way, the two electrodes can be tied to the same pin receiving the same set of instructions. Hence, the term broadcast addressing.

Generation of a broadcast addressing scheme relies on graph-based representations of electrode relationships [1]. Vertices represent electrodes, while edges represent relationships between compatible electrodes. Graph cliques can then be identified and partitioned, with the partition representing a collection of pins that can be shorted to a single driving pin (Fig. 3.1). This problem is NP-hard, but can be solved using heuristics [1]. Extensions to this basic concept include: reliability enhancement by reducing switching frequency and consequently reducing the degradation in contact angle [2], insertion of "ground vectors" for

Fig. 3.1 Broadcast addressing. (**a**) An example actuation sequence for five electrodes. (**b**) The compatibility graph, with results of the clique selection outlined in dashed lines. (**c**) The resulting broadcast-addressed, pin-mapped actuation sequence. (**d**) Electrodes are physically wired together and brought out to pins for connection to a controller

Electrode	Actuation Sequence
e_1	1 0 1 x 0 x
e_2	x 0 1 x 0 0
e_3	1 0 1 x x x
e_4	1 0 1 1 0 x
e_5	x 0 x x 1 x

(a)

(b)

Pin	Actuation Sequence
p_1	1 0 1 1 0 0
p_2	1 0 1 x 1 x

(c)

(d)

preventing residual charge [3], power consumption reduction through elimination of "redundant actuation units (RAUs)" [4]. We refer to these as toggle-aware, GV-aware, and RAU-aware pin mappers, respectively, for consistency with [5]. An optimal broadcast addressing scheme was developed in [6], achieving information-theoretical minimal pin counts. However, this scheme relies on the integration of digital logic in the biochip, which is impractical and yet to be demonstrated. Therefore we consider such DMFBs to be outside the scope of those studied in this chapter.

3.3   Security Analysis

Here, we describe the concept of tamper resistance and how it arises as a result of pin-mapped DMFBs. We then describe the threat model and derive security metrics.

3.3.1   Tamper Resistance

We claim that *the tamper resistance of a DMFB is determined by the number and distribution of don't-cares in the pin-mapped actuation sequence.* To illustrate, consider the attack in Fig. 3.2. This 15×19 DMFB is executing the InVitro 4×4 multiplexed diagnostic assay, which measures glucose (GLU), lactate (LAC), pyruvate (PYR), and glutamate (GLT) in plasma (PLA), serum (SER), saliva (SAL), and urine (URI) [7]. The attacker routes an extra plasma droplet along the red dashed line in order to alter the concentration of the sample being detected in DET3.

Normally, in a direct-addressed DMFB, such an attack is trivial to implement and would have no consequences. In a pin-mapped design, all the electrodes in yellow are unintentionally activated in the course of routing the malicious droplet. This could result in the violation of design rules, such as in MIX29, where the droplet being mixed would be unintentionally split. Subsequently, the mix operation could fail, stopping the assay progression. Or, if a randomized checkpoint system is implemented, it may detect a stray droplet where none should exist [8]. An attacker can avoid this problem by targeting pins that drive a large number of don't-cares (i.e., electrodes that are not driving droplets nor part of an interference region). Therefore, qualitatively, a tamper-resistant DMFB design will distribute don't-cares so attackers are constrained in their ability to make arbitrary droplet manipulations.

We introduce the following concepts which we will use to analyze and design tamper-resistant pin-constrained DMFBs:

Redundant units (RUs) Are don't-cares masked by either a 0 or a 1 due to the broadcast addressing scheme. This is a generalization of the redundant actuation unit (RAU)—first introduced in the power-aware pin mapper [4]—which is a don't-care that is masked by a 1. We also define a redundant deactuation unit (RDU) as a

Fig. 3.2 Tamper resistance due to pin mapping: An attacker can modify the actuation sequence to produce the malicious droplet route in dotted lines. This attack would alter the final concentration reading. The yellow electrodes are unintentionally activated in the course of the attack due to pin mapping. Unintentional splitting may occur as a result of this, making the attack evident. This pin mapping was derived from the original broadcast clique-based strategy [1]

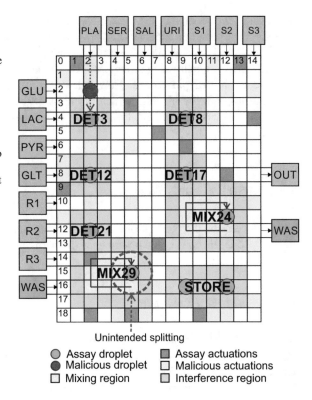

don't-care masked by a 0. For convenience, we define the function $\mathrm{RU}(p, t)$ which returns the number of redundant units associated with pin p at time-step t, as well $\mathrm{RAU}(p, t)$ and $\mathrm{RDU}(p, t)$, which return number of RAUs and RDUs, respectively. These functions map pins and time-steps to integers that are pin-dependent, i.e., $\mathrm{RU/RAU/RDU}: \mathcal{P} \times \mathcal{T} \rightarrow \mathbb{Z}$, where $\mathcal{P}$ is the set of all pins, $\mathcal{T}$ is the set of all assay time-steps. We also define $\boldsymbol{RU}$, $\boldsymbol{RAU}$, $\boldsymbol{RDU}$, as the set of all RUs, RAUs, and RDUs, respectively. A don't-care can be masked by either a 0 or a 1, but not both, so $\boldsymbol{RAU}$ and $\boldsymbol{RDU}$ are disjoint. The union of these two sets is equal to the set $\boldsymbol{RU}$. Therefore

$$\sum_{p\in\mathcal{P}}\sum_{t\in\mathcal{T}} \mathrm{RU}(p, t) = \sum_{p\in\mathcal{P}}\sum_{t\in\mathcal{T}} (\mathrm{RAU}(p, t) + \mathrm{RDU}(p, t)) \qquad (3.1)$$

Compatibility degree (CD) Measures how desirable it is to merge two electrode actuation sequences and generalizes the binary concept of compatibility. Consider two electrode actuation sequences AS_1 and AS_2, with m and n don't-cares, respectively. There are three ways in which the don't-cares can be arranged in time. In Case 1 (Fig. 3.3a), all don't-cares overlap exactly and give an attacker an opportunity for actuation tampering. In Case 2 (Fig. 3.3b), some subset of the don't-cares overlap, and in Case 3 (Fig. 3.3c), none overlap. Case 3 is the best-case

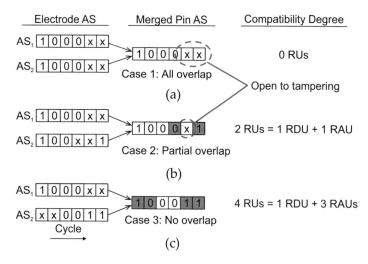

Fig. 3.3 Actuation sequence (AS) merging. (**a**) Case 1: don't-cares exactly overlap. This is undesirable as an attacker can use the resulting pin-level don't-cares to insert malicious actuations. (**b**) Case 2: partial overlap. A redundant deactivation unit (RDU) results when a broadcast 0 drives a don't-care, while a redundant activation unit (RAU) is a 1 driving a don't-care. (**c**) Case 3: no overlap. This is the most desirable from a security standpoint; an attacker is forced to affect the operation of the assay since there are no don't-cares

scenario in terms of tamper resistance; there are no exposed don't-cares, and to target a masked don't-care, an attacker will risk modifying the normal execution of the assay. Therefore, we compute the compatibility degree $CD(e_1, e_2)$ as the number of redundant units that result from the merging of two electrode actuation sequences associated with electrodes e_1 and e_2.

3.3.2   Threat Model

We assume that the *attacker* is a remote party who can access the DMFB platform through the network. Controllers for DMFBs typically incorporate a network interface either by default (e.g., when using off-the-shelf embedded computers) or by design for firmware updates and sensor data processing. The attacker is able to conduct stealthy actuation tampering attacks, i.e., extract the synthesized actuation sequences from memory, reverse-engineer them, and alter them. *The attacker does not want to be detected.* The extent of the alteration can range from simple augmentation or deletion of sequences, or can be as comprehensive as total replacement. Potential malicious actors and their motivations are discussed in [9]. The *defender* is the DMFB platform designer who wishes to ensure that any modifications to the actuation sequences are easily detectable.

3.3.3 Attack Constraints

While the threat model grants an attacker tremendous capabilities, in practice, several factors will cause attacks to become evident to the end user. Therefore, arbitrary actuation sequence modifications may not be feasible due to the following constraints:

1. *Completion time*: Assays may have completion times that are known to the end user. Relatively simple assays, e.g., sample preparation, can be assumed to execute in constant time. Such assays are commonly used as benchmarks in the DMFB literature and are studied in this chapter. More complex assays that have multiple branching points depending on intermediate results may have variable execution times [10]. Still, an end user may suspect incorrect execution if an assay completes orders of magnitude faster or slower than their experience suggests is normal.

2. *Error recovery*: DMFBs are known to be prone to several hardware faults. Cyber-physical integration has been proposed to detect and recover from errors [11]. The design of these mechanisms requires fine-tuning on the error tolerance, which may be exploitable for carrying out an attack. Furthermore, since placement of error recovery inspection points (i.e., *checkpoints*) is deterministic, a resourceful attacker could simply avoid making changes directly in critical paths [12].

3. *Intrusion detection*: Intrusion detectors can monitor parts of the biochip execution that are not actively sensed by error recovery systems. Completely deterministic detection can in theory provide 100% security, but in practice, a low overhead scheme (e.g., randomized checkpoints [8]) must be implemented.

4. *Attack surface*: We consider network-based attacks where the actuation sequence can be recovered and modified at-will. Physical fault injection attacks are possible on DMFBs, but these typically present poor localization and would be unlikely to result in a stealthy attack.

5. *Reverse engineering*: Many state-of-the-art designs for DMFBs store the actuation sequence in a format that has a one-to-one mapping between encoded bits and the biochip. Reverse engineering is thus straightforward [13]. If some mechanism were introduced to obfuscate the mapping, the attacker would not be able to make controlled changes to the assay.

6. *Unintentional fouling*: An attacker who wishes to carry out stealthy, targeted attacks will want to minimize any unintended consequences. A droplet being routed across a biochip may contaminate electrodes in its wake [14]. As these contaminations accumulate, reactions may fail or concentrations may not meet threshold criteria during checkpoints. A failure would then be detected, thwarting the attack.

3.3.4   Threat Model Refinement

We now refine the threat model according to the attack constraints:

1. *Increasing or decreasing the number of time-steps in the actuation sequence is prohibited.* This is to satisfy Constraint 1 for the non-conditional assays studied in this chapter. Even slight variations in the actuation sequence length can result in noticeable execution time differences, as DMFB actuation periods are often on the order of milliseconds (which is coarse enough to be detected by a stopwatch). Therefore, the attack can only consist of *modifications* of the actuation sequence.
2. *The number of modifications to the actuation sequence must be minimized.* This is to avoid detection by either the end user or detection by a checkpoint system (Constraints 2 and 3). In some cases, the effect of making an incremental change in the actuation sequence can be quantified; if a randomized checkpoint system is implemented, each additional change exponentially increases the probability of being detected [8].
3. *Modifications to the actuation sequence will preferentially target don't-cares.* To do otherwise would be to modify activations (1s) inserted to control droplets or deactivations (0s) inserted as part of an interference region. On pin-constrained designs, modifying a pin-level control signal will change several electrode states. Therefore, if an attacker's goal is to control a single electrode, attacking a pin may cause unintentional changes to other electrodes, potentially causing a detectable change in assay execution.

3.4   Security Metrics

We now define several security metrics to mathematically capture the notion of tamper resistance. The first, coverage, will be useful for optimization. The second, pin disturbance, is more illustrative for interpreting the results of the optimization. The third is based on randomization and provides the most direct indicator of how tamper resistance manifests itself against an attack.

3.4.1   Coverage

In Sect. 3.3, it was stated that the number and distribution of don't-cares in the pin-mapped actuation sequence determines tamper resistance. One interpretation of this is that one should mask as many don't-cares as possible.

Redundant Unit Coverage (RUC) It is defined as the proportion of electrode-level don't-cares that are masked by pin-level actuations (i.e., redundant units) over all pins and all assay time-steps. Therefore, it can be calculated as

$$RUC = \frac{\# \text{ redundant units}}{\# \text{ total don't-cares}} \qquad (3.2)$$

RUC should be maximized. In the ideal case, coverage is equal to 100%, meaning that there are absolutely no exposed don't-cares for an attacker to leverage.

Proximity Coverage Class (PCC) A variation of RUC such that only electrodes within the vicinity of a droplet are counted.

$$PCC = \frac{\# \text{ redundant units near any droplet}}{\# \text{ total don't-cares near any droplet}} \qquad (3.3)$$

Here, "near any droplet" means adjacent to the interference region along the x or y axis of a droplet. This coverage metric narrows the scope to attacks targeting functional droplets. That is, we exclude electrodes far from any functional droplets since they are unlikely to be used for manipulation attacks.

One subtle point about coverage metrics is that they are normalized. A pin mapper that generates a significant number of don't-cares can still have favorable coverage if those don't-cares are masked. That is, if we compare two pin mappers based only on the raw number of don't-cares, the one with fewer don't-cares may appear more secure. However, based on our qualitative definition of tamper resistance, this is not the case. Tamper resistance encompasses not only the fact that an attacker will try to leverage don't-cares, but also that they may attempt to tamper with functional droplets. Therefore generating a large number of redundant units can make up for the introduction of more don't-cares.

3.4.2 Pin Disturbance

We propose another set of metrics that more intuitively articulates the tamper resistance property. If we put ourselves in the perspective of an attacker, we are attempting to cause stealthy, directed changes on the biochip by altering the pin-level actuations. That is, while we want to control droplets precisely at the electrode level, our granularity of control is dictated at the pin level. Therefore, we introduce the following two metrics to measure the effect of pin-level changes:

Functional Pin Disturbance (FPD) The average number of electrodes that one would expect to disturb if a pin-level actuation or deactivation is changed. A pin-level activation (1) is used to drive one or more droplets, and will also mask some underlying don't-care electrodes. If we deactivate a pin ($1 \rightarrow 0$), then all droplets being driven by the pin will no longer be bound to the electrode. That is, the droplet will float around the biochip until it is either absorbed by another droplet or finds an empty activated electrode. This disturbance is detectable. Conversely, if we change a deactivation to an activation ($0 \rightarrow 1$), the underlying electrodes in the vicinity of a droplet will pull on the functional droplets, causing them to distort and/or split. It is desirable to maximize this quantity. Let $AU(p, t)$ be a function that returns the number of electrode-level actuations and $DU(p, t)$ be a function that returns the number of electrode-level deactivations located at pin p and time-step t.

$$\text{FPD} = \frac{1}{|\mathcal{P}||\mathcal{T}|} \sum_{p \in \mathcal{P}} \sum_{t \in \mathcal{T}} (AU(p, t) + DU(p, t)) \tag{3.4}$$

We can sum $AU(p, t)$ and $DU(p, t)$ since they are mutually exclusive on a pin-level basis.

Don't-care Pin Disturbance (DPD) The average number of electrodes that can be controlled if one changes a pin-level don't-care into an activation ($x \rightarrow 1$). Don't-cares can be freely manipulated by an attacker. A pin-level don't-care that drives multiple electrodes gives an attacker more choices for electrodes to manipulate. Let $DNT(p, t)$ return the number of electrode-level don't-cares at pin p and time t that are driven by a pin-level don't-care. It is desirable to minimize this quantity, as this gives the attacker fewer options.

$$\text{DPD} = \frac{1}{|\mathcal{P}||\mathcal{T}|} \sum_{p \in \mathcal{P}} \sum_{t \in \mathcal{T}} DNT(p, t) \tag{3.5}$$

These metrics are useful only under our stealthy attacker model. An attacker who desires to cause as much damage as possible will find that improving the disturbance level is to their benefit. Viewed another way, these metrics capture our ideal of what a tamper-resistant DMFB design should do—the actuation sequence will function correctly if undisturbed, but the slightest modification will cause a large number of easily detectable changes.

3.4.3 Probability of Detection

The previous metrics, while well-founded, may seem abstruse. A more intuitive sense of the security afforded by tamper resistance can be obtained through performing a Monte Carlo experiment: we randomly generate attacks and execute them on-chip. Some proportion of these attacks will cause interference with the assay, while others will not. The ratio of unsuccessful attacks against total attacks is the *probability of detection*, or $Pr(D)$. The randomness of the attack is both spatial and temporal. We assume that a malicious droplet will appear at any time-step and at any location on the biochip, and that it performs a random walk of random duration (subject to some maximum). At the end of the attack, we assume the droplet can then disappear. This is non-physical, but simplifies the simulation and provides an underestimate of the probability of detection; an attacker will exert even more effort to route droplets to and from the random attack.

3.5 Tamper-Resistant Pin Mapping

We propose to increase tamper resistance through pin mapping. By selectively grouping electrodes together, isolated changes to the actuation sequence will become more difficult.

3.5.1 Problem Statement

The formal problem statement is described as follows:

Input: A DMFB architecture $\mathcal{A}$ consisting of a set of electrodes $\mathcal{B}$ and a set of electrode actuation sequences $\mathcal{AS}$.

Output: A pin-constrained DMFB design assigning each electrode $b \in \mathcal{B}$ to a set of pins P, where $|\mathcal{P}| < |\mathcal{B}|$, and a set of pin-mapped actuation sequences $\mathcal{AS}_{PM}$.

Objective: Maximize the tamper resistance by maximizing redundant unit coverage.

3.5.2 Proposed Solution

The problem of grouping electrodes into pins can be modeled as a graph partitioning problem [1]. Here, we are also concerned with grouping electrodes into pins but now have imposed an additional constraint due to the desire to maximize tamper resistance. Based on our definition of compatibility degree, merging of highly compatible electrodes results in a tamper-resistant design. Therefore, the DMFB design is modeled by a graph $G = (V, E)$, where each vertex $v \in V$ represents an electrode on the DMFB array, and the set of edges E represents relationships between two compatible electrodes, similar to the original broadcast strategy. However, we now include an edge weighting function $w : E \to \mathbb{Z}$ that evaluates the compatibility degree between the two electrode actuation sequences, and a "color" function $c : V \to \mathbb{Z}$, which represents the pin assignment.

By grouping electrodes that are highly compatible, we promote solutions that increase the number of redundant units, and therefore result in a more tamper-resistant design. The grouping of high dimensional data represented by graphs is known as the graph clustering problem [15]. In particular, our specific problem of forming k number of pins out of the graph vertices associated with a distance function (i.e., "compatibility degree") is known as the minimum k-clustering problem, which has been shown to be NP-hard [16]. We therefore propose a greedy

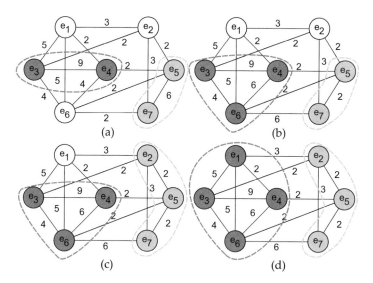

Fig. 3.4 Heuristic tamper-resistant pin mapping. (**a**) Initial phase of the heuristic algorithm. Vertices represent electrodes, weighted edges represent the compatibility degree between compatible electrodes. The highest-weighted edges are selected for the initial set. (**b**) First iteration expands the red pin by selecting a clique-compatible vertex with highest compatibility. (**c**) Second iteration expands the blue pin. (**d**) Completion of the procedure

heuristic graph-based algorithm to solve the tamper resistance optimization problem (Fig. 3.4), which proceeds as follows:

1. Initialize the graph and weights (G, w).
2. Form an initial guess p_{init} for the number of pins to form.
3. Starting from the highest-weighted edges, assign their corresponding two vertices (electrodes) to a pin by setting their color to a unique value, and skipping vertices that have already been assigned.
4. Repeat forming new pins until p_{init} pins have been formed.
5. Expand the pins by greedy iteration. For each pin, examine all neighbor vertices that are not yet assigned, and clique-compatible with the pin, add the vertex with highest compatibility, then move on to the next pin.
6. Repeat until either all electrodes are assigned or no more valid neighbors can be added.
7. Repeat the overall procedure on the remaining unassigned electrodes, and failing that, assign each an individual pin.

This is a fast heuristic algorithm that attempts to group together pins that are most compatible, with complexity $O(|V|)$ since an electrode is greedily assigned at each step. The initial guess for the pin count can be established through trial-and-error, or using knowledge of typical broadcast-addressed pin counts. The procedure is summarized in Algorithm 1.

Algorithm 1 Tamper-resistant pin mapper

Input: Set of electrodes $\mathcal{B}$, actuation sequence $\mathcal{AS}$, initial pin count p_{init}
Output: Map $c(b)$ from electrodes to pin
1: $(G = (V, E), w) \leftarrow$ initGraph($\mathcal{AS}$)
2: $c(v) \leftarrow \varnothing, \forall v \in V$
3: $sortedEdges \leftarrow$ sortDescending(E, w)
4: **for** $i \in \{0, 1, 2, \ldots, p_{init}\}$ **do**
5:     $v =$ getVertices($sortedEdges$.pop())
6:     **if** $c(v) = \varnothing$ **then**
7:         $c(v) = i$
8:     **end if**
9: **end for**
10: **while** $\exists c(v) = \varnothing$ or no colors assigned in last iteration **do**
11:     **for** $i \in \{0, 1, 2, \ldots, p_{init}\}$ **do**
12:         **for each** $\{v \in V : c(v) = i\}$ **do**
13:             $neighbors \leftarrow$ getNeighbors(v)
14:             $candidates \leftarrow \varnothing$
15:             **for** n in $neighbors$ **do**
16:                 **if** $c(n) = \varnothing$ **then**
17:                     $candidates$.add(compatibilityDegree(n, v))
18:                 **end if**
19:             **end for**
20:             $c(\max(candidates)) \leftarrow i$
21:         **end for**
22:     **end for**
23: **end while**
24: $currentPin \leftarrow p_{init} + 1$
25: **while** $\exists c(v) = \varnothing$ **do**
26:     **for each** $\{v \in V : c(v) = \varnothing\}$ **do**
27:         $c(v) = currentPin$
28:         $neighbors \leftarrow$ getNeighbors(v)
29:         $candidates \leftarrow \varnothing$
30:         **for** n in $neighbors$ **do**
31:             **if** $c(n) = \varnothing$ **then**
32:                 $candidates$.add(compatibilityDegree(n, v))
33:             **end if**
34:         **end for**
35:         $c(\max(candidates)) \leftarrow currentPin$
36:         $currentPin+ = 1$
37:     **end for**
38: **end while**
39: **return** $c(b)$

3.6 Boosting Tamper Resistance Using Indicator Droplets

Intuitively, if an actuation sequence is sparse, then we would expect that the tamper-resistant pin mapper would be unable to produce a high coverage rate. This is due to a lack of activations and deactivations to use for masking. The sparseness of an actuation sequence depends on how the bioassay has been scheduled and placed (i.e., how many concurrent operations are running), and whether routing compaction is implemented (i.e., concurrently routing droplets for the next operation phase [17]). Even if the scheduler and placer attempted to maximize concurrency, the bioassay may be constrained by critical nodes. For example, a final mix reaction may require that all previous reactions are completed first. The resulting actuation sequence would be sparse during the final cycles.

We propose to transform a bioassay such that tamper resistance is increased through the insertion of indicator droplets. If the indicator droplets are inserted properly, tampering with the actuation sequence will cause them to deviate from their expected paths. Error recovery sensors can easily detect these deviations.

It is important to note that introducing excess droplets to the biochip will degrade reliability. Droplets may accidentally pick up residues and distribute contaminants across the biochip (*cross-contamination* [18]). Excessive actuations will be introduced [19]. However, we can justify this if we are able to quantify the effect of the excess droplets and set it as a tuning parameter. This will be useful to the system designer as the amount of excess usage that is tolerable will increase as surface coatings are improved. Furthermore, if the indicator droplets are wash droplets, they will have the added benefit of cleaning the electrodes in their wake.

3.6.1   Problem Statement

We state the tamper resistance boosting problem as follows:

Input: A synthesized actuation sequence $\mathcal{AS}$.
Output: A modified actuation sequence $\mathcal{AS}_{BST}$.
Objective: Increase tamper resistance through insertion of indicator droplets while minimizing the number of excess actuations.
Constraints: Indicator droplets should obey all design rules and not interfere with functional droplet operation.

3.6.2   ILP-Based Indicator Droplet Insertion

We propose an integer linear programming (ILP) formulation to exactly solve the indicator droplet insertion problem. An electrode's state is determined by the voltage that is actually applied (the electrical state, $\mathcal{E}$), and its usage (the functional state, $\mathcal{F}$). We represent this combined state S with the notation $(\mathcal{E}, \mathcal{F})$ and code them as follows:

- *State 0* (0/0): Electrode is electrically held low (0) so as to enforce an interference region, or is deleted but part of an interference region.
- *State 1* (1/1): Electrode is electrically held high (1) so as to keep a droplet in place.

Table 3.1 ILP model electrode state notation

State S	Electrical-functional state $(\mathcal{E}, \mathcal{F})$	Description
0	$(0, 0)$	Enforcement of IR region
1	$(1, 1)$	Enforcement of droplet
2	(x, x)	Unmasked don't-care
3	$(1, x)$	Redundant activation unit
4	$(0, x)$	Redundant deactivation unit
5	(z, x)	Disconnected, not part of an IR region

- *State 2* (x/x): Electrode can be driven either high or low and is functionally a don't-care.
- *State 3* $(1/x)$: Electrode is being driven high through pin mapping, but is functionally a don't-care, i.e., a redundant activation unit.
- *State 4* $(0/x)$: Electrode is being driven low through pin mapping, but is functionally a don't-care, i.e., a redundant deactivation unit.
- *State 5* (z/x): Electrode is disconnected and not part of an interference region.

This notation is summarized in Table 3.1. Note that other combinations such as $(x/0)$ and $(1/z)$ do not arise in practice.

We now define some conventions and sets to aid the ILP formulation. The range of time-steps is $\mathcal{T} = \{0, 1, 2, \ldots, T_{\max}\}$, while the possible states are $S = \{0, 1, 2, 3, 4, 5\}$ according to Table 3.1. The coordinate convention is chosen with the origin in the top-left corner. We also add a padded area around the biochip to facilitate formulation of the ILP constraints; without doing so, we would have to specify special constraints for each of the eight edge cases: four for each corner, and four for each side. For example, the coordinate of the top-left corner on a biochip is $(0, 0)$, but becomes $(1, 1)$ when it is padded. For an $m \times n$ biochip, the range of coordinate values $C = \{0, 1, 2, \ldots, n + 1\} \times \{0, 1, 2, \ldots, m + 1\}$. After solving the model, we simply discard the border region. To simplify notation, when we sum over a single variable, we will assume it means to sum over all possible values. We also assume that the actuation sequence can be interpreted as a function $\mathcal{AS}(t, x, y) \in \{0, 1, 2\}^{T \times m \times n}$, where 0 is a deactivation, 1 is an activation, and 2 is a don't-care.

We now define our multi-objective optimization problem. The problem is to set the state of the electrodes in such a way as to minimize both the number of excess actuations and the number of unmasked don't-cares. We assume that an initial bootstrap phase is permitted, where an arbitrary number of indicator droplets can be routed onto the chip prior to the start of the assay. Introduce a binary variable $\delta_{t,x,y,s}$ which takes value 1 if electrode at coordinate (x, y) at time-step t has state s. The objective is

$$\min : \alpha \sum_t \sum_{(x,y)} \delta_{t,x,y,2} + \beta \sum_t \sum_{(x,y)} \delta_{t,x,y,1} \qquad (3.6)$$

where the first term is the sum of all non-masked don't-care terms, and the second term the sum of all electrode activations. This optimizes the tamper resistance and the reduction in reliability, respectively. α and β are weighting factors to be set by the system designer. The optimization is subject to the following constraints:

1. *Actuation sequence*: We fix electrodes associated with a functional droplet and its IR region.

$$\delta_{t,x,y,1} = 1, \quad \forall t \in \mathcal{T}, (x, y) \in C : \mathcal{AS}(t, x, y) = 1 \tag{3.7}$$

$$\delta_{t,x,y,0} = 0, \quad \forall t \in \mathcal{T}, (x, y) \in C : \mathcal{AS}(t, x, y) = 0 \tag{3.8}$$

2. *State exclusivity*: Each electrode at each time-step can only occupy a single state.

$$\sum_s \delta_{t,x,y,s} = 1, \quad \forall t \in \mathcal{T}, (x, y) \in C \tag{3.9}$$

3. *Pin mapping*: Enforcement of pin mapping requires two related constraints. The first is defined at the pin level. The sum of all $(0/x)$ states in a pin cannot be equal to the number of electrodes in the pin; to do so implies that there is no driving electrode.

$$\sum_{(x,y)\in p_i} \delta_{t,x,y,3} < |p_i|, \quad \forall t \in \mathcal{T}, \forall p_n \in \mathcal{P} \tag{3.10}$$

Similarly, the sum of all $(1/x)$ states in a pin cannot equal the number of electrodes in the pin.

$$\sum_{(x,y)\in p_i} \delta_{t,x,y,4} < |p_i|, \quad \forall t \in \mathcal{T}, \forall p_n \in \mathcal{P} \tag{3.11}$$

The second is defined at the electrode level. An electrode can be electrically high by being in state 1 OR 4. We can model this as the sum of state 1 and 4 for a particular electrode. For any two electrodes in a pin, if one electrode is being electrically driven by a 1, then the other electrode must be driven by a 1. We model this through equality of the two electrode states.

$$\delta_{t,x1,y1,1} + \delta_{t,x1,y1,4} = \delta_{t,x2,y2,1} + \delta_{t,x2,y,2,4},$$
$$\forall t \in T, \forall (x_1, y_1, x_2, y_2) \in \mathcal{Z}(p_i), \forall p_i \in \mathcal{P} \tag{3.12}$$

where $\mathcal{Z}(p_i)$ returns all transitive pairs of electrode coordinates in pin p_i. That is, if electrode A driven by a logic 1 forces electrode B to a logic 1, and electrode B forces electrode C to logic 1, then electrode A will also force electrode C to logic 1. This constraint is unnecessary for pins with only a single electrode.

4. *Removed electrodes*: Some electrodes are not used in the course of protocol's execution. The pin mapper will "delete" these electrodes by not connecting

any pins to them. The system designer can then choose not to fabricate these electrodes. For modeling purposes, we assume that these electrodes still exist, but allow them to only be in state 0 or 5 by fixing all other state variables to 0. Let $\mathcal{R}$ denote the set of coordinates corresponding to electrodes removed during pin mapping.

$$\sum_t \delta_{t,x,y,s} = 0, \quad \forall k \in \{1, 2, 3, 4\}, \forall (x, y) \in \mathcal{R} \tag{3.13}$$

For all other electrodes, we can disallow setting them to the disconnected state (5).

$$\sum_t \delta_{t,x,y,5} = 0, \quad \forall (x, y) \in C \setminus \mathcal{R} \tag{3.14}$$

5. *Interference regions*: If an electrode has an indicator droplet (i.e., has state 1), it must keep an interference region of 0 deactivations around it. Functional assay droplets are derived from synthesis and already have appropriate IR enforcement. They also must be able to merge and split. Applying IR constraints to functional droplets will cause infeasibility. To avoid this, let $\mathcal{W}(t, x, y)$ be a function returning the set of coordinates surrounding an electrode at (t, x, y). We define the following for all don't-cares in the actuation sequence:

$$8 \cdot \delta_{t',x',y',1} \leq \sum_{(x,y) \in \mathcal{W}(t',x',y')} \delta_{t',x,y,0},$$
$$\forall t' \in \mathcal{T}, (x', y') \in C : \mathcal{AS}(t', x', y') = 2 \tag{3.15}$$

6. *Continuity*: Droplets must maintain their presence between time-steps. Let $N(x, y)$ return the set with coordinates (x, y) and its direct North/South/East-/West neighbors.

$$\delta_{t',x',y',1} \leq \sum_{(x,y) \in N(x',y')} \delta_{t+1,x,y,1},$$
$$\forall t' \in \mathcal{T}, (x', y') \in C : \mathcal{AS}(t, x, y) = 2 \tag{3.16}$$

7. *Dispense/Waste ports*: Indicator droplets can only be introduced or consumed at dispense and waste ports, respectively. Activations for all other electrodes must be disallowed. Let $\mathcal{I}$ be the set of coordinates for all dispense and waste ports. The constraint is not imposed at time-step 0 to generate an initial bootstrap solution.

$$\delta_{t,x,y,1} = 0,$$
$$1 \leq t \leq T_{\max}, \forall (x, y) \in C \setminus \mathcal{I} \tag{3.17}$$

3.6.3 Iterative ILP-Based Sliding Window Approximation

The exact ILP-based formulation (Table 3.2) is optimal, but is clearly intractable for all but the simplest DMFBs. This mainly stems from the time complexity of the model used; all time-steps are considered in forming the optimal solution, which be in the tens of thousands. To overcome this, we propose several approximations and an ILP-based sliding window algorithm.

We approximate the minimization of unmasked don't-cares by introducing the concept of *impact*, which is measured for each electrode and time-step of an actuation sequence. An electrode has high impact if, when we place a droplet on it, a large number of don't-cares are masked. Define a helper function insertDroplet$(\mathcal{AS}, t, x, y)$ which returns a state vector $s \in \mathcal{S}^{(m+1)\times(n+1)}$ representing the biochip at time t if we insert a droplet at position (x, y). Also define a function countRUs(s) which counts the number of redundant units in a state vector. Then we simply proceed to insert test droplets and count redundant units for each electrode and each time-step, collecting the results into a vector $w_{t,x,y}$. Therefore the procedure takes $O(\mathcal{XYT})$. Calculation of impact is summarized in Algorithm 2.

The simplified ILP model depends on preprocessing the actuation sequence to find valid locations for placing an indicator droplet. Let $\mathcal{V}(t, x, y)$ be a function taking on value 1 if electrode (x, y) at time-step t is a don't-care or masked actuation $(1/x)$ with sufficient interference region clearance, and 0 otherwise. The procedure for preprocessing is summarized in Algorithm 3, and also has worst-case complexity $O(\mathcal{XYT})$. We then derive our approximate ILP model as follows. Define the binary variable $\delta_{t,x,y}$ to take value 1 if we are to add an indicator droplet at coordinate

Table 3.2 Symbols used in ILP model for solving the indicator droplet placement problem

Symbol	Elements	Description
$\mathcal{T}$	$\{0, 1, 2, \ldots, T_{max}\}$	Set of all time-steps
$\mathcal{X}$	$\{0, 1, 2, \ldots, n + 1\}$	Set of all padded x-coordinates
$\mathcal{Y}$	$\{0, 1, 2, \ldots, m + 1\}$	Set of all padded y-coordinates
C	$\mathcal{X} \times \mathcal{Y}$	Set of all padded biochip coordinates
$\mathcal{P}$	$\{p_0, p_1, p_2, \ldots, p_{max}\}$	Set of all pins
p_i	$\{(x_0, y_0), (x_1, y_1), .., (x_{|p_i|}, y_{|p_i|})\}$	Set of coordinates of electrodes in pin i
$\mathcal{Z}(p_i)$	$\{(x_0, y_0, x_1, y_1), (x_1, y_1, x_2, y_2), \ldots\}$	Function returning set of all transitive pairs of electrode coordinates in p_i
$\mathcal{R}$	$\{(x_0, y_0), (x_1, y_1), .., (x_{|\mathcal{R}|}, y_{|\mathcal{R}|})\}$	Set of coordinates of deleted electrodes
$\mathcal{W}(t, x, y)$	$\{(x_0, y_0), \ldots, (x_7, y_7)\}$	Function returning set of surrounding neighbor coordinates
$\mathcal{N}(x, y)$	$\{(x_0, y_0), \ldots, (x_3, y_3)\}$	Function returning set of N/S/E/W neighbors
$\mathcal{I}$	$\{(x_0, y_0), (x_1, y_1), .., (x_{|\mathcal{I}|}, y_{|\mathcal{I}|})\}$	Coordinates of all IO ports

---

Algorithm 2 Electrode impact

Input: Actuation sequence $\mathcal{AS}$
Output: Impact vector $w_{t,x,y}$
1: $w_{t,x,y} \leftarrow \varnothing, \forall v \in \mathcal{T}, (x, y) \in C$
2: **for** $\forall t \in \mathcal{T}, (x, y) \in C \setminus \mathcal{R}$ **do**
3:     testActuation $\leftarrow$ insertDroplet(AS, t, x, y)
4:     $w_{t,x,y} \leftarrow$ countRUs(testActuation)
5: **end for**
6: **return** $w_{t,x,y}$

Algorithm 3 Preprocessing for valid electrodes

Input: Actuation sequence $\mathcal{AS}(t, x, y)$
Output: Map $\mathcal{V}(t, x, y)$ indicating validity of electrode (x, y) at time-step t
1: $\mathcal{V}(t, x, y) \leftarrow \varnothing$
2: **for** $\forall t \in \mathcal{T}$ **do**
3:     **for** each pin-connected electrode $\forall (x, y) \in C \setminus \mathcal{R}$ **do**
4:         **if** $\mathcal{AS}(t, x, y)$ is don't-care or $(1/x)$
             and all neighbors in $\mathcal{W}(t, x, y)$ equal don't-care or $(0/x)$ **then**
5:             $\mathcal{V}(t, x, y) \leftarrow 1$
6:         **end if**
7:     **end for**
8: **end for**
9: **return** $\mathcal{V}(t, x, y)$

(x, y) and time-step t, 0 otherwise. The objective is to maximize impact by placing a single indicator droplet.

$$\max : \sum_{t} \sum_{(x,y)} w_{t,x,y} \delta_{t,x,y} \tag{3.18}$$

Subject to the following constraints:

1. *Valid electrodes*: Due to the presence of functional droplets and pin mapping, only some electrodes are eligible to add droplets. We fix invalid locations to 0.

$$\delta_{t,x,y} = 0, \quad \forall t \in \mathcal{T}, (x, y) \in C : \mathcal{V}(t, x, y) = 1 \tag{3.19}$$

2. *Single droplet routing*: We can only add a single droplet in each time-step. This is a self-imposed constraint so as to speed up computation. Additional droplets are added during the iteration phase.

$$\sum_{(x,y)} \delta_{t,x,y} \leq 1, \quad \forall t \in \mathcal{T} \tag{3.20}$$

3. *Continuity*: This is defined identically as the original ILP formulation in Eq. (3.16).

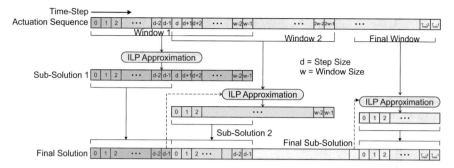

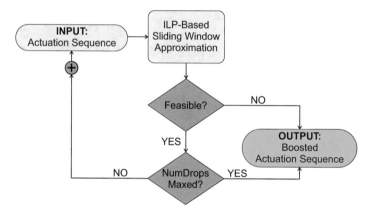

Fig. 3.5 ILP-based sliding window approximation algorithm. We subdivide the problem into smaller manageable chunks, and solve them using the simplified ILP approximation algorithm. We construct the final solution by truncating and appending the sub-solutions. The final window may be smaller than the other windows if the actuation sequence length is not an integer multiple of the window size

Fig. 3.6 Iterative ILP-based sliding window approximation algorithm. Indicators are successively added to the actuation sequence until either the model is infeasible or the maximum number of indicators have been reached

Despite our simplifications, a large number of time-steps will still lead to intractable models. To overcome this, we can use the ILP approximation as the basis of an iterative sliding window algorithm. Define a window width of w time-steps and a step size of d time-steps, and iterate by solving the ILP approximation for the first w time-steps. We take the first d solutions, and then solve the next window of time-steps offset by d. We repeat and keep adding solutions until the overall problem is solved. The procedure is summarized in Fig. 3.5. Each iteration of this algorithm yields a boosted actuation sequence with one additional indicator droplet. We can iteratively repeat the procedure on the boosted actuation sequence until no more can be added (Fig. 3.6).

3.7   Experimental Results

We evaluated our pin mapper against the broadcast addressing [1], RAU-aware [4], ground vector-aware (GV-aware) [3], and toggle-aware [2] pin mappers using four benchmark assays: PCR, InVitro 4×4, Protein, and Protein Split 5. The benchmark simulation data was generated with the open-source MFStaticSim tool [5] using a 15×19 DMFB array, virtual topology placer, and Roy maze router. The tamper-resistant pin mapper was implemented in MATLAB. The ILP-based sliding window algorithm was implemented in Python 3.6.4 using Gurobi 7.5 as our ILP solver. We used an Intel Core i7-8700 3.2 GHz machine with 16 GB RAM. The window size was set to 100 time-steps and the step size was set to 50 time-steps. The most complex assay, ProteinSplit5, took more than 1 day to route the initial indicator droplet. Subsequent indicator droplets routed within a few hours.

3.7.1   Baseline Performance

We summarize the baseline performance in Fig. 3.7, which shows pin counts, switching toggles (SW) measured in thousands as an inversely related indicator of power and reliability, RUC and PCC measured in percentages. We see that tamper-resistant pin mapper achieves, on average, 39.6% higher RUC than the next-best prior work, which is the broadcast clique pin mapper. When we restrict attacks to the proximity coverage class (PCC), coverage can be as high as 73.9% for some assays using the tamper-resistant pin mapper. Most of the pin mappers achieve coverage rates of less than 50% across all assays. At the same time, this work's pin counts are modestly increased while switching activity is increased but remains on the same order. We note that the pin counts are still reasonable for implementation on a PCB. Additionally, compatibility in the RAU-aware pin mapper was enabled between actuation sequences with opposite polarities, with the reasoning that an inverter could be inserted in the biochip. While this does save on pins, most commercially available DMFBs and prototypes today feature completely passive biochips. Therefore, the pin-count savings may not be realizable in practice. Therefore, we conclude that the proposed algorithm achieves its goal and is able to produce a quantifiably more tamper-resistant pin-constrained DMFB.

3.7.2   Performance with Indicator Droplets

Performance for the boosted assays is shown in Fig. 3.8. As expected, inserting indicator droplets incrementally increases the security performance. Each pin mapper exhibits a similar rate of change. Since the tamper-resistant pin mapper starts out with better metrics, it continues to beat the other pin mappers as indicator droplets are added. In our experimental data, pin mappers for the PCR assay

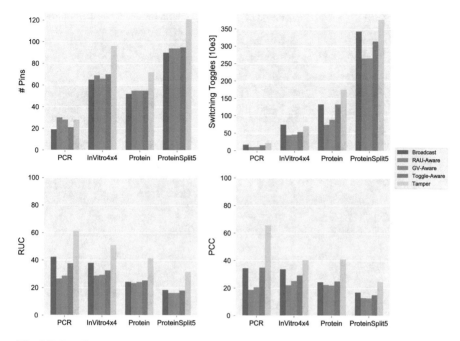

Fig. 3.7 Baseline comparison of the tamper-resistant pin mapper against broadcast [1], RAU-aware [4], GV-aware [3], and toggle-aware [2] pin mappers. The number of pins is increased, but still within an acceptable range for integration on a PCB. Switching toggles are increased. This work achieves, on average, 39.6% higher RUC than the next-best prior work, the broadcast clique pin mapper

typically deleted around 200 out of 285 electrodes, meaning that there is very little degree of freedom for droplet routing. This is exhibited in the saturation of the curves after two or three indicator droplets; this shows that the solver fails to find ways to insert indicator droplets. The ability to plot the change in switching toggles with indicator droplets allows one to select an appropriate number of indicator droplets for a given piece of hardware.

3.7.3   Probability of Detection

Probability of detection is shown in Fig. 3.9. We performed 1000 Monte Carlo experiments for each assay and pin mapper combination, setting the maximum random walk length to 10. Increasing the random walk only disadvantages the attacker, as this gives more time-steps for a detection system to discover the malicious droplet (see Chap. 4). We observe that the tamper-resistant pin mapper achieves better baseline probability of detection than all other pin mappers. As indicator droplets are added, the performance gap decreases, although the general trend is maintained. This demonstrates that arbitrary attacks can be detected with

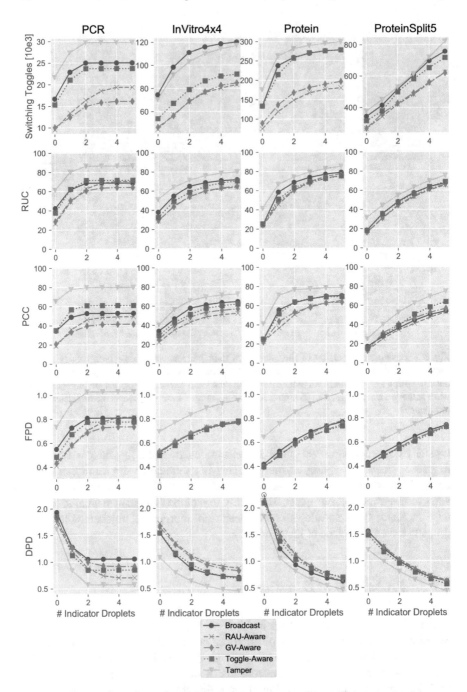

Fig. 3.8 Security metrics vs. indicator droplets. Successively adding indicator droplets improves the security metrics but also increases excess switching toggles. In some pin mappers, the PCR assay saturates after a few indicator droplets are added

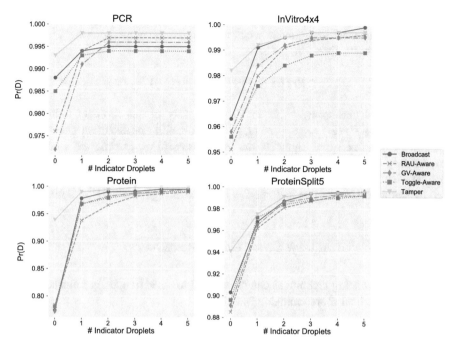

Fig. 3.9 Probability of detection vs. indicator droplets. The tamper-resistant pin mapper outperforms other pin mappers, while successively adding indicator droplets improves $Pr(D)$ for all pin mappers

high probability, approaching 1 in many cases, and that the tamper-resistant pin mapper is advantageous if we require that no indicator droplets are to be added. The only exception to the trend is in the InVitro 4×4 assay, where after the fifth indicator droplet is added, the broadcast addressing scheme actually outperforms the tamper-resistant pin mapper.

3.8 Discussion

The basis of tamper resistance is the idea that certain actions by an attacker are detectable. It is conceivable that a clever adversary could design the attack in such a way that it meets all the correctness criteria set by execution checkpoints. In this case, it is recommended that a randomized checkpoint system be utilized [8, 20]. Randomized checkpoints give a probability of evasion that is parameterized by attack length L. We denote this quantity with hat notation as $Pr(\hat{E})$. We may quantify the interaction of randomized checkpoints with tamper resistance by converting the probability of detection metric into a probability of evasion as $Pr(E) = 1 - Pr(D)$. The overall system's probability of evasion, denoted by boldface as $Pr(E)$, is the probability that the attacker evades both the randomized checkpoints and evades a collision with the unaltered actuation sequence. These are independent events, therefore we have:

$$Pr(E) = Pr(\hat{E})(1 - Pr(D)) \qquad (3.21)$$

Of course, we may convert this into a probability of detection by finding the complement as $Pr(D) = 1 - Pr(E)$. The probability of detection encountered in practice can be quite high. On the PCR assay, if we assume the worst-case probability of evasion from a randomized checkpoint system that monitors 10% of the electrodes, we have $Pr(\hat{E}) = 0.90$ [8]. Tamper-resistant pin mapping with no indicator droplets results in $Pr(D) = 0.993$, which gives $Pr(D) = 0.9937$.

3.8.1 Comparison with Other Countermeasures

A number of existing techniques can be adopted to make tampering infeasible, but with some practical shortcomings:

- *Encryption* is often the first defense that comes to mind for both protecting the privacy and integrity of an actuation sequences. However, encryption in hardware systems is not foolproof; poor key management schemes have often led to insecure systems. Also, at some point, the actuation sequence *must* be decrypted to be executed on hardware. This is the same reason that full-disk encryption is not secure against malware; it only defends against unauthorized access when the device is not active. Moving the encryption to hardware bears significant implementation costs. Furthermore, encryption implementations may be susceptible to side-channel attacks [21]. Proper use of encryption is non-trivial to achieve.
- *Cryptographic hashes* can provide a unique signature for an actuation sequence pattern, but will fail if the signature is just as susceptible to attack as the actuation sequence—which is often the case. This situation bears similarity to software download hashes, which are a good idea for checking against random transmission errors, but poorly suited when an adversarial threat model is adopted.
- *Code signing* is a commonly employed method for ensuring the integrity of firmware. In fact, the main reason for Stuxnet's success was the lack of code signing [22]. However, this simple countermeasure can fail; major corporations have been known to lose control of their private signing keys [23–25].

In contrast, incorporating security measures in the pin mapping phase of the DMFB design flow is highly advantageous as no extra circuitry or control hardware is required. Our approach has modest pin count overhead, which is acceptable since the PCB layer count can form a substantial portion of overall system cost anyway [26]. Furthermore, this is a hardware-based technique that is not susceptible to attacks that take advantage of network-enabled controllers. And by inserting indicator droplets, we substantially increase tamper resistance using an essentially free resource: water.

3.8.2 Extensions for Electrode Weighting

On a biochip, certain electrodes may be more critical for the operation of the assay than others. For example, some electrodes may be used for mix and detection operations, while others are only used exclusively for routing of droplets. We can make these critical electrodes more tamper-resistant by weighting them appropriately. For the pin mapping phase, the weighting can be introduced by changing the definition of compatibility degree (Sect. 3.3.1). For the indicator droplet problem, weighting ($w_{t,x,y}$) is already included in the simplified ILP formulation (Eq. (3.18)). Changing the definition of electrode impact is all that is required. The optimal ILP formulation can be modified to incorporate weighting by changing the first term in Eq. (3.6) to be a weighted linear combination of $\delta_{t,x,y,2}$. While it is simple to augment the developed techniques for electrode weighting, it is likely that some experimentation will be required to find the appropriate weights.

3.9 Summary and Conclusion

This chapter presented the first study of DMFB pin mappers as a tamper resistance mechanism. The restriction on droplet movements imposed by pin mappers simultaneously lowers an attacker's ability to arbitrarily route droplets, and causes undesirable side-effects on other droplets existing on the chip. We introduced the redundant unit coverage security metric to describe the masking of don't-cares. Experimental results show that existing pin mapping algorithms, while optimizing for reliability and power consumption, lead to poor tamper resistance. A new pin mapping algorithm was proposed to increase masking effects, and ILP-based techniques were developed to insert indicator droplets to boost tamper resistance even further with modest, quantifiable pin count and switching overhead.

Tamper-resistant pin mapping is a simple design-time technique to harden a DMFB design against actuation tampering attacks. It is hardware-based and is therefore intrinsic to the operation of the DMFB, mitigating the effects of an attack should high-level security mechanisms fail. It comes at no cost to the designer, as a pin mapping algorithm must be chosen regardless of security concerns. There is, however, a fundamental trade-off in control complexity and switching toggles required for a given level of security. This trade-off is quantifiable and thus tunable by the system designer.

The proposed methodology ensures high coverage rates as a post-synthesis processing step. While optimizing for tamper resistance during the high-level synthesis phases could in theory provide better performance, we believe it is more productive to leverage all of the rich literature on DMFB synthesis. Furthermore, methodologies that do not subdivide the synthesis tasks often have long runtimes despite being based on heuristics [27].

This work does have some limitations that can be addressed as future work. We have considered pin mappers only as a post-synthesis processing step. Therefore, the optimization problem is highly constrained. This explains why 100% coverage is not achievable. It is conceivable that even better tamper resistance properties could be achieved if other aspects of the synthesis flow, such as placement and routing, were optimized for tamper resistance. The experimental results show that some overhead is required to optimize for tamper resistance. This can be adjusted through trial-and-error, but it would be preferable to fine-tune the balance as a multi-objective optimization problem.

We note that the assays studied in this work are static and do not feature conditional execution [28]. The concept of tamper resistance through pin mapping can be adapted to conditional assays, but would require some care in its implementation. The conditionally executed actuation sequences must be synthesized in such a way that it is compatible with a given pin mapping. The original broadcast addressing scheme provides one such method for adding additional operations to an existing pin-mapped design through appropriate scheduling [1]. Additionally, the security metrics would need to be redefined since a reference actuation sequence is not available as a reference for correctness. It should be possible to quantify the "effort" required to move a droplet from a random source to a random destination.

Other future work includes investigating methods to deter DoS attacks, and to extend the work to be applicable to emerging micro-electrode-dot-array (MEDA) biochips [29, 30].

References

1. Y. Zhao, T. Xu, K. Chakrabarty, Broadcast electrode-addressing and scheduling methods for pin-constrained digital microfluidic biochips. IEEE Trans. Comput. Aided Des. Integr. Circuits Syst. **30**(7), 986–999 (2011)
2. S.-T. Yu, S.-H. Yeh, T.-Y. Ho, Reliability-driven chip-level design for high-frequency digital microfluidic biochips. IEEE Trans. Comput. Aided Des. Integr. Circuits Syst. **34**(4), 529–539 (2015)
3. T.-W. Huang, T.-Y. Ho, K. Chakrabarty, Reliability-oriented broadcast electrode-addressing for pin-constrained digital microfluidic biochips, in *Proceedings IEEE/ACM International Conference Computer-Aided Design* (2011), pp. 448–455
4. T.-W. Huang, H.-Y. Su, T.-Y. Ho, Progressive network-flow based power-aware broadcast addressing for pin-constrained digital microfluidic biochips, in *Proceedings IEEE/ACM Design Automation Conference* (2011), pp. 741–746
5. D. Grissom, C. Curtis, S. Windh, C. Phung, N. Kumar, Z. Zimmerman, O. Kenneth, J. McDaniel, N. Liao, P. Brisk, An open-source compiler and PCB synthesis tool for digital microfluidic biochips. Integr. VLSI J. **51**, 169–193 (2015)
6. T. A. Dinh, S. Yamashita, T.-Y. Ho, An optimal pin-count design with logic optimization for digital microfluidic biochips. IEEE Trans. Comput. Aided Des. Integr. Circuits Syst. **34**(4), 629–641 (2015)

7. F. Su, K. Chakrabarty, High-level synthesis of digital microfluidic biochips. ACM J. Emerg. Technol. Comput. Syst. **3**(4), 1 (2008)

8. J. Tang, M. Ibrahim, K. Chakrabarty, R. Karri, Secure randomized checkpointing for digital microfluidic biochips. IEEE Trans. Comput. Aided Des. Integr. Circuits Syst. **37**, 1119–1132 (2018)

9. S.S. Ali, M. Ibrahim, O. Sinanoglu, K. Chakrabarty, R. Karri, Security assessment of cyberphysical digital microfluidic biochips. IEEE/ACM Trans. Comput. Biol. Bioinform. **13**(3), 445–458 (2016)

10. M. Ibrahim, K. Chakrabarty, K. Scott, Synthesis of cyberphysical digital-microfluidic biochips for real-time quantitative analysis. IEEE Trans. Comput. Aided Des. Integr. Circuits Syst. **36**(5), 733–746 (2017)

11. Y. Luo, K. Chakrabarty, T.-Y. Ho, Error recovery in cyberphysical digital microfluidic biochips. IEEE Trans. Comput. Aided Des. Integr. Circuits Syst. **32**(1), 59–72 (2013)

12. Y. Zhao, T. Xu, K. Chakrabarty, Integrated control-path design and error recovery in the synthesis of digital microfluidic lab-on-chip. ACM J. Emerg. Technol. Comput. Syst. **6**(3), 11 (2010)

13. H. Chen, S. Potluri, F. Koushanfar, BioChipWork: reverse engineering of microfluidic biochips, in *Proceedings IEEE International Conference Computer Design (Newton, MA)* (2017) pp. 9–16

14. H. Yao, Q. Wang, Y. Shen, T.-Y. Ho, Y. Cai, Integrated functional and washing routing optimization for cross-contamination removal in digital microfluidic biochips. IEEE Trans. Comput. Aided Des. Integr. Circuits Syst. **35**(8), 1283–1296 (2016)

15. S.E. Schaeffer, Graph clustering. Comput. Sci. Rev. **1**(1), 27–64 (2007)

16. T.F. Gonzalez, Clustering to minimize the maximum intercluster distance. Theor. Comput. Sci. **38**, 293–306 (1985)

17. T.-W. Huang, T.-Y. Ho, A fast routability- and performance-driven droplet routing algorithm for digital microfluidic biochips, in *Proceedings IEEE International Conference Computer Design* (2009), pp. 445–450.

18. Y. Zhao, K. Chakrabarty, Cross-contamination avoidance for droplet routing in digital microfluidic biochips. IEEE Trans. Comput. Aided Des. Integr. Circuits Syst. **31**(6), 817–830 (2012)

19. J.K. Park, S.J. Lee, K.H. Kang, Fast and reliable droplet transport on single-plate electrowetting on dielectrics using nonfloating switching method. Biomicrofluidics **4**(2), 024102 (2010)

20. J. Tang, M. Ibrahim, K. Chakrabarty, Randomized checkpoints: a practical defense for cyberphysical microfluidic systems. IEEE Des. Test **36**(1), 5–13 (2018)

21. D. Agrawal, B. Archambeault, J.R. Rao, P. Rohatgi, The EM side-channel(s), in *Proceedings International. Workshop Cryptographic Hardware Embedded System* (Springer, Berlin, 2002), pp. 29–45.

22. R. Langner, Stuxnet: dissecting a cyberwarfare weapon. IEEE Secur. Priv. **9**(3), 49–51 (2011)

23. T. Mendelsohn, Secure boot snafu: Microsoft leaks backdoor key, firmware flung wide open (2016)

24. D. Goodin, In blunder threatening windows users, d-link publishes code-signing key (2015)

25. B. Schneier, Sony PlayStation 3 master key leaked (2012)

26. D.T. Grissom, J. McDaniel, P. Brisk, A low-cost field-programmable pin-constrained digital microfluidic biochip. IEEE Trans. Comput.-Aided Des. Integr. Circuits Syst. **33**(11), 1657–1670 (2014)

27. D.T. Grissom, P. Brisk, Fast online synthesis of digital microfluidic biochips. IEEE Trans. Comput. Aided Des. Integr. Circuits Syst. **33**(3), 356–369 (2014)

28. D. Grissom, C. Curtis, P. Brisk, Interpreting assays with control flow on digital microfluidic biochips. ACM J. Emerg. Technol. Comput. Syst. **10**(3), 24 (2014)

29. G. Wang, D. Teng, Y.-T. Lai, Y.-W. Lu, Y. Ho, C.-Y. Lee, Field-programmable lab-on-a-chip based on microelectrode dot array architecture. IET Nanobiotechnol. **8**(3), 163–171 (2013)

30. G. Wang, D. Teng, S.-K. Fan, Digital microfluidic operations on micro-electrode dot array architecture. IET Nanobiotechnol. **5**(4), 152–160 (2011)

Chapter 4
Detection: Randomizing Checkpoints on Cyberphysical Digital Microfluidic Biochips

4.1 Introduction

The basic principle of detecting and correcting anomalous behavior underlies all work on DMFB error recovery using cyberphysical integration [1]. Therefore, taken in a security context, error recovery systems provide protection from actuation tampering attacks. While a system that monitors the entire biochip at every electrode for all time is guaranteed to detect all anomalies, such an implementation would impose severe processing and memory constraints on DMFB designs where the goal is to miniaturize and keep costs low. Indeed, the realization of the "lab-on-a-chip" depends on making any additional security features as lightweight as possible. Randomization permits fast probing of the biochip state while causing uncertainty for an attacker.

We develop the idea of a randomized checkpoint system by defining security metrics and techniques for improving performance through static checkpoint maps. Performance trade-offs associated with static and random checkpoints are also discussed. We provide detailed classification of attack models and demonstrate the feasibility of our techniques with case studies on assays implemented in typical DMFB hardware. As we will see, such a defense mechanism is simple to implement while providing high assurances of integrity. When coupled with a prevention mechanism, performance can be increased even more.

4.2 Threat Model

We consider a cyberphysical DMFB under an actuation tampering attack. That is, we assume that a malicious adversary is able to modify the low-level actuation sequences of the DMFB hardware to an extent that she may purposefully dispense, route, and mix droplets. This could occur through the insertion of hardware

© Springer Nature Switzerland AG 2020
J. Tang et al., *Secure and Trustworthy Cyberphysical Microfluidic Biochips*,
https://doi.org/10.1007/978-3-030-18163-5_4

Trojans, or through modification of the control software. Control software may be compromised through a network connection originally integrated for software updates or convenient downloading of assay specifications [2]. The attacker can read out, reverse engineer, and modify the actuation sequences stored in the DMFB controller. They are aware of any deterministic error detection schemes and can thus craft their attack to avoid them. The operator of the DMFB is presumed to be trustworthy, and that there is no tampering with the physical aspects of the system such as the loading of samples/reagents and the imaging system. They are unable to breach the security coprocessor due to the lack of a network interface.

We make no assumptions about the architecture of the DMFB. The techniques described in this chapter are general and can be applied to both general-purpose programmable DMFBs and application-specific DMFBs. However, there may be more use for random checkpoints on a general-purpose biochip, since an attacker would have access to more resources to carry out an attack. Error recovery may or may not be implemented. If it happens to be present on the biochip under consideration, an attacker is assumed to be able to predict and evade their location since they are placed using deterministic algorithms [1].

Example: Denial-of-Service

One potential threat is illustrated in Fig. 4.1. This example shows the final execution cycles of a polymerase chain reaction (PCR) assay. The PCR assay is used in DNA amplification and has been studied extensively in the DMFB literature. At clock cycle i, a malicious droplet is dispensed from the AmpliTaq DNA polymerase port to be routed to mix module $M1$. Higher concentrations of AmpliTaq increase production of nonspecific products, lowering the quality of the assay output [3–5]. Either the DMFB error detection scheme will detect the wrong concentration at the output of this mix stage or this altered droplet will be allowed to propagate through

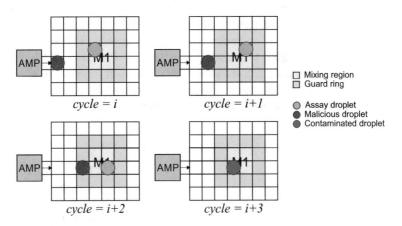

Fig. 4.1 Example malicious route. Droplets can be dispensed to foul/contaminate a good droplet in a mix module. The target droplet concentrations can be altered to either cause failure of the assay (denial-of-service) or to report incorrect measurements

the assay if no error recovery is implemented. In either case, the result of the attack is either a denial-of-service or output alteration/contamination.

The consequence of output contamination/alteration is that an assay result may be interpreted as accurately reflecting reality. This can be dangerous in cases where the assay is used to perform some measurement or test [6], for example, in vitro glucose measurement. If a user's glucose measurement is inaccurate, the wrong dosage of insulin may be administered which could lead to overdose. The result of a DoS attack is that the DMFB is not able to perform its intended function, causing inconvenience while wasting samples, reagents, and money. But more insidiously, a DoS attack, if not detected as a DoS attack, may trick error correction to believe that a hardware fault has occurred. Electrodes may be marked as faulty when they are still functional, causing the DMFB hardware to have reduced fault-tolerance and shorter operating lifespan.

4.2.1 Attack Modeling

All practical malicious modifications require the movement of droplets from a *source* to a *target* on the DMFB. Examples of sources include dispense ports, waste reservoirs, and backup reservoirs. Examples of targets include output ports, backup reservoirs, mix modules, and droplets in transit. It is conceivable that an adversary could mount an attack that does not alter the result, such as dispensing extra reagents into unused electrodes, but the focus in this chapter remains on attacks that change the assay result.

Hence, we model the class of attacks that can be formulated as a misrouting problem between a source and a target. It should be noted that not all (source, target) combinations result in a meaningful attack. For example, routing a wash droplet into an output port would be easily detected as a fault since it bears no resemblance to the desired output droplet. This observation will save some time in analyzing and evaluating the proposed defense system in Sect. 4.3. Table 4.1 enumerates the typical resources in a DMFB platform and classifies them as potential sources or targets for a malicious droplet.

Table 4.1 Classification of resources in DMFB as potential sources and targets for an actuation tampering attack

Resource	Source	Target
Dispense port	✓	
Output port		✓
Waste reservoir	✓	
Backup reservoir	✓	✓
Mix module	✓	✓
Droplets in transit	✓	✓

4.2.2 Attack Classification

Malicious modification of a DMFB actuation sequence can be classified according to the degree of modification as follows:

- *Level 1: Bit flip.* A single bit in the actuation sequence is modified. Such an attack can be achieved through physically inducing errors in the hardware, similar to reported fault injection attacks in cryptographic hardware [7].
- *Level 2: Sequence modification/insertion/deletion.* N bits of the actuation sequence can either be modified, inserted, or deleted. An intelligent adversary would be able to manipulate droplets in such a way that most of the assay proceeds normally.
- *Level 3: Complete substitution.* The most extreme attack is to completely replace the correct actuation sequence with an alternate sequence. The result of such an attack is likely to be noticed, since error recovery mechanisms can detect the deviation from specification. Additionally, large deviations in processing time could be detected by the DMFB operator.

This chapter addresses level 2 attacks as they have the most potential to induce harm while being difficult to detect.

4.3 Randomized Checkpoints

The proposed randomized checkpoint system consists of security coprocessor, which is physically isolated from the DMFB controller (Fig. 4.2). Attackers are unable to breach the security coprocessor due to the lack of a network interface. The coprocessor is able to selectively probe the status of droplets on the biochip and compare them to the assay specification. The coprocessor should have a separate physical indicator to alert the DMFB operator when an anomaly is detected.

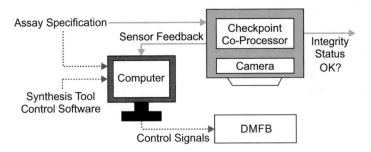

Fig. 4.2 DMFB secure coprocessor implementation schematic. Solid lines indicate signals assumed to be trustworthy, while dotted lines are susceptible to attack

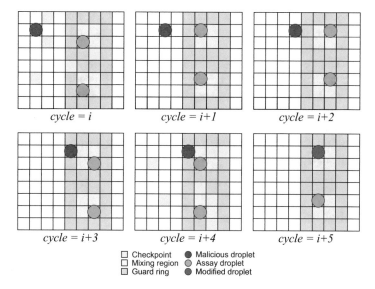

Fig. 4.3 Example assay execution with random checkpoints applied at cycle i and $i + 4$. In this case, $s = 72, k = 5, L = 5$. This example is a realization of the random sampling times. Assuming $c = 0.33$, then $Pr(E) = 0.89$

Checkpointing, or the monitoring of an assay's progression, is a technique for ensuring the integrity of an assay. Given unlimited resources, the most secure action is to record the entire assay at a high sample rate and analyze the log for anomalous behavior. This soon becomes cost prohibitive; the computing and memory requirements for analyzing the data from a high-resolution camera will quickly erode any low-cost benefits of using DMFB technology. Instead of full security, we seek a solution with probabilistic guarantees that the assay has not been tampered with. This is achieved by sampling the biochip's execution randomly in both time and space. The proposed system works as follows (Fig. 4.3):

1. *Determination of which electrode to examine*: The system randomly chooses an electrode according to a uniform distribution over the electrodes. That is, assuming a DMFB with s number of electrodes, we assign an index $j \in \{0, 1, \ldots, s-1\}$ to each electrode according to some predefined convention (e.g., left-to-right, top-to-bottom) and define a random variable J which represents the outcome of randomly selecting an electrode. The probability mass function (PMF) is defined as $p_J(j) = 1/s$ for $j \in \{0, 1, \ldots, s-1\}$.
2. *State extraction*: The controller focuses the imaging system on the electrode and runs a correlation algorithm against a template image to extract the state of the droplet [1]. In general the state may include volume and concentration, but in this chapter we assume that it is sufficient to extract only the presence or absence of a droplet.

3. *Comparison with specification*: The controller will then compare the state of electrode j against the specification stored in memory, and signal an error if they do not match.
4. *Repeat*: The previously chosen electrode is marked as chosen, and a new electrode is chosen from the remaining pool. The PMF is now $p_J(j) = 1/(s - |X|)$ $j \in \{0, 1, \ldots, s - 1\} \setminus X$, where X is the set of electrodes already chosen. The system repeats this process up until a number k defined by the system designer. Since the DMFB is physically limited by the fundamental actuation frequency, all the inspection events within a time-step can be considered to be occurring simultaneously.

4.4   Probability of Evasion

We quantify the effectiveness of our defense system through the probability of evasion, $Pr(E)$. If we assume that the system can fully monitor the entire biochip at every time-step, then $Pr(E) = 0$. On the contrary, a DMFB without any detection capabilities will have $Pr(E) = 1$. It is desirable to minimize $Pr(E)$.

It is not possible to exactly anticipate what an attack will entail. Therefore, we model the attack in terms of a "malicious droplet." That is, a malicious adversary has to route droplets on-chip as part of their attack. We can establish a lower bound on $Pr(E)$ by examining one such malicious droplet, and assuming that it exists over L time-steps. By doing so, we ignore the particular details of the malicious route and can instead analyze potential attacks in terms of distances between sources and targets.

Let E be the event that a malicious droplet evades detection for the lifetime of the droplet that executes over L total cycles. L may be much less than the lifetime of the assay. E_i is the event that a malicious droplet evades detection for the i-th execution cycle, and F_i is the event that the i-th cycle is sampled and G_i is the event that the i-th cycle's checkpoints intersect with the malicious droplet. If each cycle's checkpoints are chosen independently, and the events F_i and G_i are independent, then the evasion event is equivalent to the event that the cycle is not sampled, or the cycle is sampled and the set of checkpoints does not monitor the droplet.

$$Pr(E_i) = Pr(\overline{F_i} \cup (F_i \cap \overline{G_i})) = Pr(\overline{F_i}) + Pr(F_i) \cdot Pr(\overline{G_i}) \tag{4.1}$$

$$Pr(E) = Pr(E_1 \cap E_2 \cap \cdots \cap E_L) = \prod_{i=1}^{L} Pr(E_i) \tag{4.2}$$

The probability that a malicious droplet does not intersect with any checkpoints is the complement of the ratio of active checkpoints k at that time over the number of total electrodes s. We have

$$Pr(\overline{G_i}) = 1 - \frac{k}{s} \tag{4.3}$$

The ratio k/s is called the *electrode coverage ratio*. Then we define the probability of sampling any execution cycle using some constant c as

$$Pr(F_i) = c \tag{4.4}$$

This constant is a design parameter that can be adjusted in software. Therefore, the probability of evasion can be expressed as

$$Pr(E) = \prod_{i=1}^{L} \left((1-c) + c\left(1 - \frac{k}{s}\right) \right) = \left(1 - \frac{ck}{s}\right)^L \tag{4.5}$$

The parameter s is a constant determined by the size of the DMFB array. The parameters c and k should be maximized in order to minimize the likelihood of evasion, subject to the computational and imaging capabilities of the DMFB platform. Note that the system has no control over the route an attacker may take, and that this model does not assume any particular malicious route. The only assumption on the attack is that it exists for a certain number of cycles L, and that the probability of evasion has an exponential dependence on L (Fig. 4.4). This key observation leads to the possibility of decreasing $Pr(E)$ through influencing the routability of malicious as a result of judicious placement of static checkpoints.

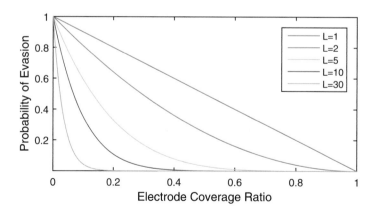

Fig. 4.4 The probability of evasion decreases as the checkpoint coverage ratio, k/s, increases. Here $c = 1$, meaning checkpoints are monitored on every execution cycle. Increasing L as a parameter shows exponential advantage, illustrating the intuitive notion that the longer a malicious droplet exists on an assay, the more likely it is to be detected

4.5 Biased Probability Distributions

The analysis presented in Sect. 4.3 assumes that each electrode is chosen to be sampled independently and with uniform probability. The system can bias the distribution, since it is plausible that some electrodes are more useful for an attacker than others. However, an intelligent adversary would be able to use this information to avoid electrodes that are more likely to be detected. It is not immediately clear if there is any benefit to biasing because of the threat model. Before we show how to reconcile this conflict, we introduce some new definitions and generalizations.

4.5.1 Biased PMF

We can generalize the probability of evasion to consider non-uniform distributions over the electrodes and define our PMF as

$$p_J(j) = 1/s + b(j) \tag{4.6}$$

where $b(j)$ is a bias term. That is, we describe the PMF in terms of deviations from the uniform distribution. Note that $\sum p_J = 1 \Rightarrow \sum b(j) = 0$, and $b(j) \in (-1/s, 1 - 1/s)$. In the uniform case, $b(j) = 0$. For convenience, we will notate $b^0(j)$ when we know that $b(j)$ will evaluate to 0 for some j, and similarly notate $b^+(j)$ and $b^-(j)$ when we know there will be a positive or negative bias, respectively.

4.5.2 Generalized Probability of Evasion

Equation (4.5) is a function of several variables. We introduce the notation $P_{Ei}(j, k, b(j))$ to mean the probability of the event E_i during cycle i as a function of the currently occupied electrode indexed by j, with k number of random checkpoints per cycle and some bias function b evaluated at j. Generalizing Eq. (4.2) we write the product not only over the cycle i but also the attack *Path*, where a *Path* is defined as an ordered set of ordered pairs (i, j) indicating the time-step and location of a path. That is, a *Path* $\subset (\mathbb{N} \times \mathbb{N})^L$

$$Pr(E) = \prod_{i,j \in Path} P_{Ei}(j, k, b(j)) \tag{4.7}$$

Using b^0 simplifies to the analysis in the previous section, while $k = 1$ simplifies P_{Ei} equal to $1 - p_J(j)$. Evaluating $P_{Ei}(j, k, b(j))$ in general is difficult, since it is equivalent to the probability of selecting k combinations out of $s - 1$ electrodes not occupied by the droplet. However, we do know that an electrode with higher bias is less likely for an attacker to evade detection than an electrode with lower bias. That is, if $b(\alpha) > b(\beta)$, then $P_{Ei}(\alpha, k, b(\alpha)) < P_{Ei}(\beta, k, b(\beta))$.

4.5.3 Decomposition of Probability of Evasion

Now assume an arbitrary DMFB array with uniform distribution. If we perturb the distribution on some electrode α such that $b(\alpha) = \delta$ for $\delta \in (-1/s, 1 - 1/s)$, we must also distribute $-\delta$ among one or more of the other electrodes in order to satisfy $\sum b(j) = 0$. The probability that a malicious route evades detection was given in Eq. (4.7). We can break this equation into three components according to bias. Denote some arbitrary bias $b^-(j) < b^0(j) < b^+(j)$. We rewrite Eq. (4.7) in terms of these three biases as

$$Pr(E) = \prod_{x \in X} P_{Ei}(x, k, b^-(x))$$
$$\times \prod_{y \in Y} P_{Ei}(y, k, b^0(y)) \qquad (4.8)$$
$$\times \prod_{z \in Z} P_{Ei}(z, k, b^+(z))$$

where X is the set of electrodes with negative bias, Y is the set of electrodes with no bias, and Z is the set of electrodes with positive bias, and $Path = X \cup Y \cup Z$. Equation (4.8) can be thought of as the multiplication of $|X| + |Y| + |Z|$ number of probability terms, where each probability term in X is less than every term in Y is less than every term in Z. In certain cases, this decomposition can facilitate the relative comparison of performance between two routes.

4.5.4 Security of Biased Distributions

The security of a given bias distribution is determined by the worst-case performance. The worst-case performance with a uniform distribution is determined by the shortest attack route. We denote P_{Emin} as the worst-case (shortest-path) probability of evasion for the uniform distribution. Any probability of evasion for a biased distribution P_E^* should not exceed this limit. That is,

$$P_E^* \leq P_{Emin} \qquad (4.9)$$

It is clear that a route with one or more negatively biased electrodes has higher $Pr(E)$ than one that is unbiased. Similarly, a route with one or more positively

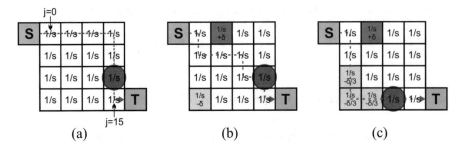

Fig. 4.5 Effect of biasing electrodes. (**a**) Representation of a uniformly distributed electrode arrangement. (**b**) Biasing with $+\delta$ on electrode $j = 1$ and $-\delta$ on electrode $j = 12$. (**c**) Biasing with compensating delta spread out over multiple electrodes. $b(j) = +\delta$ ($j = 1$), $-\delta/3$ ($j = \{8, 12, 13\}$), $1/s$ (otherwise). Electrodes in set Y are white, X in light gray, and Z in dark gray

biased electrodes has lower $Pr(E)$. Due to the fact that any positive bias has to be compensated by a negative bias on another electrode, it is conceivable that there are bias schemes that do not provide any net benefit. Since each electrode can only be negatively biased by a maximum of $-\delta = 1/s$, attempting to apply more than $+\delta = 1/s$ bias to one or more electrodes means that more than one electrode is adversely affected. Furthermore, if on a particular DMFB architecture, every electrode is part of more than one minimum-length route, that means applying a compensating bias will always have negative consequences on another route.

Figure 4.5a illustrates a uniform probability distribution over a typical general-purpose DMFB source–target configuration. If a positive bias is applied in an attempt to improve the odds of catching the malicious route in red in Fig. 4.5b, then a negative bias needs to be applied elsewhere. This can be done in several ways. Figure 4.5b shows the negative bias applied to an alternate route. Figure 4.5c shows the bias being distributed over three electrodes, forming an easier path for an attacker to take. Neither case is preferable, since the negative bias causes the condition in Eq. (4.9) to be violated.

In general, most electrodes on a general-purpose DMFB grid are part of more than one minimum-length route. Therefore, for security reasons and for simplicity of system design, the recommended distribution for most DMFB architectures is the uniform distribution.

4.6 Static Checkpoint Placement

While randomized checkpoints are the foundation of the detection scheme, the addition of static checkpoints can increase the overall effectiveness. Static checkpoints by themselves provide weak security guarantees; it relies on their location being kept secret. Under our proposed threat model, locations of these static checkpoints are known to the attacker. An attacker with such knowledge is best served by avoiding the static checkpoints. The judicious placement of static checkpoints can influence the type of routes an attacker will take.

4.6.1   Problem Statement

We represent a general-purpose DMFB array as the integer grid $\mathbb{N} \times \mathbb{N}$, with a set of sources $S \subset \mathbb{N} \times \mathbb{N}$ of cardinality s and a set of targets $T \subset \mathbb{N} \times \mathbb{N}$ of cardinality d. The problem is to find the smallest set of static checkpoints (or obstacles) $K \subset \mathbb{N} \times \mathbb{N}$ of cardinality k such that the smallest routing length between every source and target is maximized. In other words, we seek solutions to the problem

$$\underset{K}{\text{argmax}} \; \min(\text{allRoutes}(S, T, K)) \tag{4.10}$$

where allRoutes(S, T, K) is the set of all possible routes between every source and target that does not collide with an obstacle in K, and min$(\ldots)$ returns the smallest route of this set. The function to be maximized is difficult to solve because it considers all feasible route lengths rather than just the Manhattan distance or L_1 norm. For an $m \times n$ grid, there are $\binom{m \times n}{k}$ ways to place k checkpoints. Even on a relatively small DMFB platform with 120 electrodes and 20 checkpoints that amounts to over 2^{74} combinations to choose from. Furthermore, the computation of the set of feasible routes is non-trivial ($O(mn)$ for maze type router [8]).

4.6.2   Minimal Provably Secure Placement

The problem in Eq. (4.10) describes an optimal arrangement of static checkpoints. However, we can relax this requirement and still provide security guarantees by defining a *minimal provably secure* checkpoint map as a set K such that

$$\min(\text{allRoutes}(S, T, K)) > \min(\text{allRoutes}(S, T, \varnothing)) \tag{4.11}$$

In other words, we seek any set of checkpoints that causes the smallest possible route between the sources and targets to be greater than the smallest possible route without any static checkpoints. On a general-purpose biochip with no obstacles, a droplet can be routed with a Manhattan length route. We propose a solution set K to this problem using Hadlock's theorem [9] and concepts from graph theory.

Hadlock's Theorem: *A path α from P to Q has length $d_M(P, Q) + 2 \cdot d$, where d is the detour number of α with respect to Q and $d_M(P, Q) = |x_Q - x_P| + |y_Q - y_P|$ is the Manhattan distance between P and Q.*

Since the detour number d is a number greater than or equal to 0, the path length is minimized when $d = 0$. Therefore the shortest route length is equal to the Manhattan distance. This somewhat obvious statement provides a baseline by which to measure the performance of any obstruction placement algorithm. Furthermore, the fact that no detours are permitted allows us to make a simplification to the problem.

To solve for a set K that increases the route length, we can represent an array of DMFB electrodes as a directed graph $G = (V, E)$, where each vertex represents an electrode and each edge represents a possible path. Associated with each vertex is an (x, y)-coordinate describing its location on the DMFB array, where the origin is taken to be the upper-leftmost electrode. On an unmodified electrode grid, a droplet is free to move in any direction, so each forward edge has a matching return edge (Fig. 4.6a, b).

Since no detours are permitted, we can convert the directed graph G into a modified graph G^* where all detour edges are removed (Fig. 4.6c). For a target node T, a detour edge is defined as any edge connecting a two nodes A and B, where $(y_T - y_A) < (y_T - y_B)$ or $(x_T - x_A) < (x_T - x_B)$. That is, G^* represents all possible minimal length routes. If we desire to eliminate all possible minimal length routes, the problem is to choose where to break the graph. Breaking the graph is equivalent to placing static checkpoints on the boundary of the cut, as an attacker cannot cross the cut. We seek solutions where the cut has the fewest number of edges, since the number of edges translates directly into static checkpoints that must be scanned. The problem of inserting static checkpoints such that a longer route is forced is equivalent to solving the minimum $s - t$ cut problem with each edge being assigned a uniform flow. We may compute such a solution using any of the available algorithms in the literature [10, 11]. This result is general and can apply to any DMFB architecture. However, the simple symmetric structure of general-purpose DMFBs leads to a predictable solution for optimal placement: electrodes directly adjacent to sources and targets should be monitored (Fig. 4.6c).

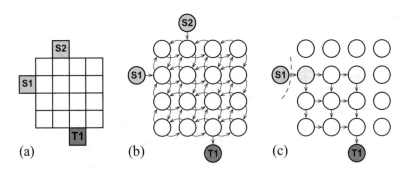

Fig. 4.6 Graph-based static checkpoint placement. (**a**) Sample DMFB architecture with two sources and one target. (**b**) Graph representation $G = (V, E)$ of the DMFB. (**c**) Transformed graph representation $G^* = (V^*, E^*)$ for determination of static checkpoint placement between S2 and T1. Note that grid structures give rise to simple solutions where checkpoints should be placed close to either source or target

To create a checkpoint map that secures the entire biochip, it suffices to enumerate all combinations of sources and targets, generate each individual checkpoint map using the $s - t$ min-cut formulation, and superimpose them. The resulting superposition can be interpreted as a matrix where each individual cell represents the number of static checkpoints that were added. The resulting map thus presents a ranking of electrodes in terms of how many attacks it can be used to carry out. In a constrained environment, the designer may choose the highest ranked electrodes. While this checkpoint map is provably secure, it is not optimal in that another checkpoint map may exist which achieves the same performance with fewer checkpoints.

4.6.3 Heuristic Placement

Static checkpoint placement can be approximated by a heuristic algorithm, since it has been observed that symmetric general-purpose DMFBs admit predictable solutions. Furthermore, we have some intuitive notion of where it might be useful to place static checkpoints; we can prevent many attacks by placing a single checkpoint at droplet sources. The idea is to rank each DMFB electrode in terms of usefulness to an attacker for routing a malicious droplet, and to select electrodes from this list up to a number defined by hardware limitations.

We give an approximate ranking of electrodes by enumerating all useful combinations of sources and sinks, and form a rectangle containing the farthest edges of these resources. A matrix representation of the DMFB is constructed where the area corresponding to this bounding box is filled with ones. Such a matrix is called an *electrodeWeight* matrix. For a source s and target t, the corresponding matrix is *electrodeWeight(s,t)*. Electrodes closer to the source are given higher weights according to the intuition that security issues are easier to stop at the source. Figure 4.7 illustrates a simplified example with one source dispense port ($D1$) and two destination mix modules ($M1$, $M2$).

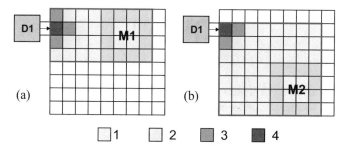

Fig. 4.7 Heuristic static checkpoint placement. (**a**) Rectangle used to approximate usefulness of electrodes in harboring a malicious route for dispense port 1 and mix module 1. (**b**) Rectangle for approximation between dispense port 1 and mix module 2

The corresponding matrices for the example in Fig. 4.7 are:

$$electrodeWeight(1, 1) = \begin{bmatrix} 3\,2\,1\,1\,1\,1\,1\,0 \\ 4\,3\,2\,1\,1\,1\,1\,0 \\ 3\,2\,1\,1\,1\,1\,1\,0 \\ 2\,1\,1\,1\,1\,1\,1\,0 \\ 0\,0\,0\,0\,0\,0\,0\,0 \\ 0\,0\,0\,0\,0\,0\,0\,0 \\ 0\,0\,0\,0\,0\,0\,0\,0 \\ 0\,0\,0\,0\,0\,0\,0\,0 \end{bmatrix} \tag{4.12}$$

$$electrodeWeight(1, 2) = \begin{bmatrix} 0\,0\,0\,0\,0\,0\,0\,0 \\ 4\,3\,2\,1\,1\,1\,1\,0 \\ 3\,2\,1\,1\,1\,1\,1\,0 \\ 2\,1\,1\,1\,1\,1\,1\,0 \\ 1\,1\,1\,1\,1\,1\,1\,0 \\ 1\,1\,1\,1\,1\,1\,1\,0 \\ 1\,1\,1\,1\,1\,1\,1\,0 \\ 1\,1\,1\,1\,1\,1\,1\,0 \end{bmatrix} \tag{4.13}$$

The weighting for assay as a whole, denoted as *arrayWeight*, is calculated by adding each combination of source and sink electrode weights as follows:

$$arrayWeight = \sum_i \sum_j electrodeWeight(source(i), target(j)) \tag{4.14}$$

$$= electrodeWeight(1, 1) + electrodeWeight(1, 2)$$

$$= \begin{bmatrix} 3\,2\,1\,1\,1\,1\,1\,0 \\ 8\,6\,4\,2\,2\,2\,2\,0 \\ 6\,4\,2\,2\,2\,2\,2\,0 \\ 4\,1\,1\,1\,1\,1\,1\,0 \\ 1\,1\,1\,1\,1\,1\,1\,0 \\ 1\,1\,1\,1\,1\,1\,1\,0 \\ 1\,1\,1\,1\,1\,1\,1\,0 \end{bmatrix} \tag{4.15}$$

The resultant ranking matrix is intuitive. On a simple DMFB architecture, the best location to insert static checkpoints is immediately in front of a malicious droplet source. If the result of the heuristic algorithm produces identical rankings, the checkpoint placer should randomly select between them. Note that the result cannot guarantee that any routes are actually longer or more difficult than in the unaltered case, in contrast to the algorithm presented in the previous section. The electrode weighting approach is summarized in Algorithm 1.

Algorithm 1 Electrode ranking

Input: DMFB architecture $\mathcal{A}$, set of sources S and targets T
Output: Matrix ranking each electrode, $arrayWeight$
1: $arrayWeight \leftarrow 0$
2: **for each** combination $s \in S$ and $t \in T$ **do**
3:    **if** timestep of $t \le$ timestep of s **then**
4:       $arrayWeight \leftarrow arrayWeight + electrodeWeight(s,t)$
5:    **end if**
6: **end for**
7: **return** $arrayWeight$

4.6.4   Temporal Randomization of Static Checkpoints

A static checkpoint can be monitored with some probability v instead of at every cycle. This will increase the probability of evasion, but there may be certain scenarios where this is an acceptable trade-off to decrease average power consumption. Let q be the number of static checkpoints, and Q be the number of static checkpoints on the malicious route. Then the probability of evasion can be modeled as

$$Pr(E) = (1 - v)^Q \left(1 - \frac{ck}{s - q}\right)^{L-Q} \tag{4.16}$$

If the static checkpoints are monitored 100% of the time, the probability of evasion is exactly zero, unless the malicious route does not cross any static checkpoints ($Q = 0$ leads to Eq. (4.5)). When a malicious route crosses a static checkpoint instead of an electrode that is under random sampling, the following inequality must hold for there to be a net performance gain:

$$(1 - v) \le \left(1 - \frac{ck}{s - q}\right) \tag{4.17}$$

Rearranging, we find

$$v \ge c\left(\frac{k}{s - q}\right) \tag{4.18}$$

which can be interpreted as a tuning requirement. Recall that v and c are constants to be tuned by the system designer. Since the $k/(s - q)$ term is less than or equal to 1, v can be less than c while still lowering $Pr(E)$. Therefore, static checkpoints take fewer resources to implement than randomized checkpoints for the same level of security while potentially increasing the difficulty of an attacker to minimize their route length.

4.6.5 Security of Checkpoint-Based Error Recovery

The concept of a checkpoint can be generalized in order to account for checkpoints inserted by error recovery techniques. Thus, the security provided by the error recovery system can be evaluated. We represent a general checkpoint as an ordered 7-tuple

$$C_j = \langle x(j), y(j), i(j), vol_{low}(j), vol_{high}(j), conc_{low}(j), conc_{high}(i) \rangle \qquad (4.19)$$

where x and y are the coordinates that the detection takes place, i is the actuation cycle that detection occurs, $vol_{low}(j)$ and $vol_{high}(j)$ define a valid interval of droplet volumes, and $conc_{low}(j), conc_{high}(j)$ define a valid interval of concentrations. These interval specifications can be set to *don't-care* values simply by setting the low value to 0, and the high value to largest value perceptible by the imaging system.

We define an arrangement of checkpoints M as a set of k randomized checkpoints

$$M(i) = \{C_1, C_2, \ldots, C_k\} \qquad (4.20)$$

where each x, y coordinate is chosen uniformly from the possible electrodes of the DMFB array, without replacement. These checkpoints are all active at the same cycle i. Testing for the presence of a droplet is specified by setting $vol_{low}(j)$ to the smallest droplet volume that can be manipulated by the DMFB hardware and $vol_{high}(j)$ to its maximum value. Testing for the absence of a droplet is specified by setting both $vol_{low}(j)$ and $vol_{high}(j)$ to zero. We do not consider the concentration in the checkpoint system, so concentration is set to *don't-care*. Finally, a randomized checkpoint system C_{rand} can be defined as a set of arrangements M, where each cycle i is selected by performing a *Bernoulli(c)* trial for each cycle of the assay execution. A checkpoint arrangement is added to the set for each success. The arrangement should be generated on-the-fly and with a high quality random number generator.

Based on this interpretation, it can be seen that the proposed security mechanism provides some level of error detection, while error detection provides some measure of security. That is, they both can determine when the behavior of a DMFB system deviates from its intended operation. Correct attribution of an error is difficult in practice. For error recovery, it is difficult to infer that a malicious adversary was the cause since faulty hardware could potentially produce the same result. On the other hand, in checkpoints used for detecting a malicious adversary, faulty hardware can lead to false positives.

Attributing a fault to either an attack or hardware failure can be done to some extent by analyzing the observed droplet behavior. For instance, if a droplet is detected at a specific point before any droplets specified by the actuation sequence have had a chance to reach it, an attack has almost certainly occurred since no failure mode is likely to have caused this behavior. Faults such as stuck droplets are more ambiguous; an attacker could easily induce this failure, but if historical reliability data for the DMFB platform indicates that stuck droplets are highly probable, then the end user could reasonably conclude that the fault was caused by hardware failure. In general though, correct attribution of observed faults is non-trivial and end users may be required to consider extraneous factors such as how connected the device is, or whether poor access control policies are implemented.

4.7   Realistic System Constraints

As a reminder, the motivation for a randomized checkpoint system is to lower the amount of resources required to monitor assay execution. The probability of evasion was found to decrease monotonically with the number of checkpoints monitored in a given time-step. Therefore the best strategy is always to use as many checkpoints as possible. Before discussing the following case studies, it is instructive to investigate constraints on k imposed by realistic DMFB controller platforms.

The CCD camera provides sensor data in the form of an array of raw pixel values. Droplet presence, volume, concentration can be extracted from these pixels through a pattern matching algorithm. A practical matching algorithm consists of focusing the CCD camera at specific points on the biochip, and calculating the correlation between the captured image and a template image [1]. Figure 4.8 shows a C implementation of the correlation algorithm described in [1]. The reference template image is passed as an argument in array x, while the cropped sub-image from the CCD is passed in array y. A larger correlation value means the two images

```c
#include <math.h>
#define T_SIZE 625
float cor(int x[T_SIZE], int y[T_SIZE]) {
  int num=0, den_x=0, den_y=0, sum_x=0, sum_y=0;
  int xavg, yavg;
  for(int j=0; j<T_SIZE; j++){
    sum_x += x[j];
    sum_y += x[j];
  }
  xavg = sum_x / T_SIZE;
  yavg = sum_y / T_SIZE;

  for(int i=0; i<T_SIZE; i++){
    num += (x[i]-xavg)*(y[i]-yavg);
    den_x += (x[i]-xavg)*(x[i]-xavg);
    den_y += (y[i]-yavg)*(y[i]-yavg);
  }
  return (float) num / sqrt((float) den_x*den_y);
}
```

Fig. 4.8 C implementation of the droplet correlation algorithm

are strongly related. Assuming a 25×25 template image and compiling the code to target the STM32F2x7 series of mid-range ARM Cortex-M3 microcontrollers, this code can require more than 28,500 clock cycles to execute. Operating the microcontroller at 120 MHz would thus require 238 μs to examine a single droplet. A typical operating frequency for a DMFB is 100 Hz; therefore, no more than 42 droplets can be examined in a single actuation cycle.

Based on this information, the following case studies will assume the checkpoint coprocessor is limited to examining 20 checkpoints in each actuation cycle (i.e., $k + q = 20$) in order to leave some computing cycles for system overhead. In both cases, the probability of evasion will be evaluated with system parameters $c = 0.5$ and $v = 0.5$, which sets the probability that either static or random checkpoints are examined equal to a fair coin flip. The case where $c = 1$ and $v = 0.5$ is also investigated. The system should be capable of executing with both parameters equal to 1, but it may be desirable to reduce dynamic power consumption by throttling the amount of checking.

4.8 Case Studies

The following case studies demonstrate the type of performance that can be expected from the randomized checkpoint system. Each biochip architecture and assay are drawn from the research literature and realistic constraints are used to extract the performance metrics. The attacks are simulated using an open-source DMFB synthesis tool [12, 13] modified to incorporate our checkpoint techniques. Attacker routing is simulated using the Lee routing algorithm [8], which implements the attacker's optimal strategy of minimizing the malicious route. Monte Carlo simulation is used to validate the analysis, with the probability of evasion being given by the complement of the ratio of successful detections to number of attempted trials.

4.8.1 Polymerase Chain Reaction

The polymerase chain reaction is used for the amplification of DNA and has been demonstrated on a number of microfluidic platforms. We study the PCR protocol executing on a general-purpose DMFB architecture under DoS attack. The goal of the attacker is to route a KCl droplet from the dispense port to mix module $M4$ (Fig. 4.9), as excess KCl concentration inhibits PCR [14]. The target module $M4$ is bound by the synthesis tool to mix KCl with Tris-HCl.

Random Checkpoints Only The route taken by our malicious software router is shown in Fig. 4.9a. It takes 8 execution cycles to reach its destination, which is minimal since it is equal to the Manhattan distance. Note that it is possible to route

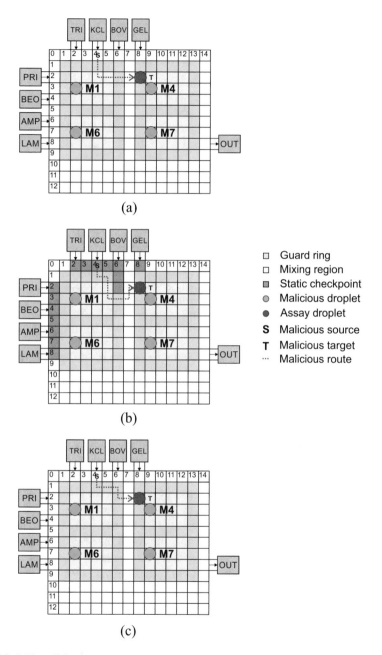

Fig. 4.9 PCR malicious route case study. (**a**) Random checkpoints only. Malicious route from KCl port targets M4 mix module. No obstacles mean that the adversary is free to route with minimal distance, minimizing probability of evasion. (**b**) Random and static checkpoints. The minimum Manhattan distance is no longer achievable, decreasing the probability of evasion. (**c**) Random checkpoints with error recovery. Malicious route is redirected to avoid mixing region $M1$ due to error recovery checkpoints. While the original route is no longer achievable, there still exists a minimal path from the source to target

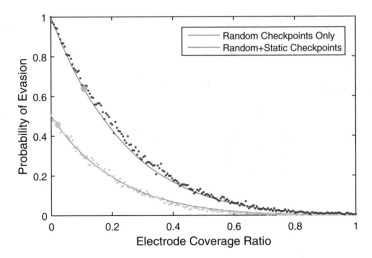

Fig. 4.10 Probability of evasion vs. electrode coverage ratio for a minimal length route and a route with static checkpoints attempting to dispense KCl into $M4$. Probability that a given cycle is monitored was set to 50% for both random and static checkpoints. The two large dots indicate data points from the PCR case study, where the number of checkpoints is limited to 20

the droplet through mix module 1 if the timing is chosen carefully; there is a small window of execution where the droplet being mixed in $M1$ is too far to be affected by the malicious droplet. Figure 4.10 illustrates how $Pr(E)$ varies as a function of the electrode coverage ratio for the given route. With the given constraint of $k = 20$, the electrode coverage ratio is 10.3%. This route yields $Pr(E) = 0.66$.

Random and Static Checkpoints We assume the adversary is able to learn about the static checkpoints and thus is able to route around them. We fix $q = 16$ and use the heuristic algorithm in Sect. 4.6 to place the static checkpoints. The result is that all the dispense ports are covered. The provably secure placer gives the same result. With q set and total checkpoints limited to 20, $k = 4$. With $v = 0.5$, the inequality in Eq. (4.18) is satisfied, so the optimal strategy for the adversary is to route around the static checkpoints if possible. Figure 4.9b illustrates the placement of the top 16 static checkpoints and the path chosen by the router. Note that the malicious route is forced to cross one static checkpoint ($Q = 1$) and then takes 9 cycles to reach the target ($L = 9$). Figure 4.10 shows $Pr(E)$ for this longer route as being lower than the original route for all electrode coverage ratios. Blocking off the dispense port provides a tremendous advantage. In this case, the electrode coverage ratio is 2.2% and $Pr(E) = 0.45$. Note that if v had been set to 1, this attack would have been detected immediately. Setting c to 1 would further increase system performance (Fig. 4.11).

Random Checkpoints with Error Recovery Checkpoints Now we investigate how error recovery checkpoints interact with the system. A malicious route attempting to cross through an error recovery checkpoint would immediately trigger the

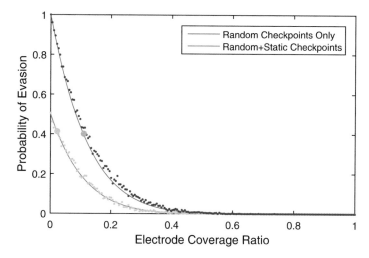

Fig. 4.11 Performance improves when the probability of monitoring random checkpoints at each cycle is increased to 100%. Probability of monitoring static checkpoints is 50%. The two large dots indicate data points from the PCR case study, where the number of checkpoints is limited to 20. $Pr(E)$ drops to 0.40 at 10.3% coverage ratio for random checkpoints only

error recovery process. Error recovery checkpoints are placed at critical junctions in the protocol specification such as mix operations. Assuming the attacker's goal has not changed, the route must now move around the region defined by $M1$, which is being actively monitored (Fig. 4.9c). The malicious droplet joins the droplet being mixed in $M4$. This is detectable depending on the how the error recovery system is set up. If the error thresholds are not set properly, changing the concentration of KCl this way is conceivable. The route length is the same as the case in Fig. 4.1. Thus the error recovery mechanism provides no advantage in terms of detecting malicious droplets in transit, and $Pr(E)$ is exactly the same as in the case without error recovery. We note that error recovery mechanisms can provide an advantage in indirectly detecting attacks, if the attackers are not careful in staying within error thresholds.

4.8.2   Commercial 3-Plex Immunoassay

Figure 4.12 shows a commercial 3-plex immunoassay [15] DMFB with 1038 electrodes, excluding the dispense electrodes. This design is an application-specific biochip designed for acute myocardial infarction diagnosis. Despite the non-reconfigurable nature of this device, an attacker may still modify the actuation sequence to stall or introduce droplets into the reaction region. Droplets are dispensed from the routing region, processed in the reaction region and then sensed in the detection region. Error recovery is not implemented on this biochip.

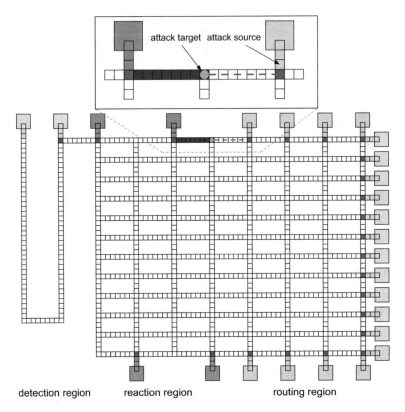

Fig. 4.12 Application-specific DMFB architecture for carrying out an n-plex immunoassay. The static checkpoint placer targets the outputs of the dispense ports (red electrodes). The heuristic algorithm results are identical when the top 23 electrodes are chosen

The static checkpoint placement algorithms were run on this architecture, resulting in the placement maps indicated in Fig. 4.12. The difference between the results of the optimal vs. heuristic placement algorithms are negligible. In both cases, it can be seen that they capture the intuitive notions of which electrodes are more important to monitor, which is near the dispense ports. We model an attacker who is interested in diluting the sample droplets so as to alter the final detected result. This type of attack has serious implications for quality of patient care. The malicious route is illustrated in dashed lines in the detail of Fig. 4.12; it attempts to dispense a reagent into one of the linear mixing regions.

Random Checkpoints Only The malicious route takes 9 steps to get from the reagent dispense port to the linear reaction chamber. This route length is minimal, and it is difficult for an attacker to choose an alternate route since this application-specific architecture provides only a single pathway for each sample to be processed. With the constraint of $k = 20$, the electrode coverage ratio is limited to 1.93%, which sets $Pr(E) = 0.92$ (Fig. 4.13). When $c = 1$ this decreases to $Pr(E) = 0.86$

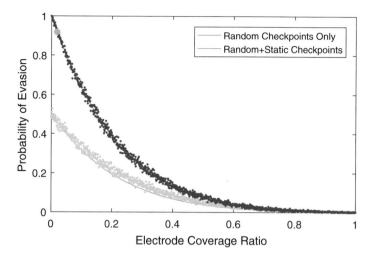

Fig. 4.13 Probability of evasion vs. electrode coverage ratio for an attack causing dilution of the reaction chamber. The probability of sampling both random and static checkpoints is set to 50% ($c, v = 0.5$). Solid lines show analytic results, dotted lines are simulation data. Solid dots indicate operating points given the assumption of a maximum of 20 checkpoints

(Fig. 4.14). The size of the DMFB is poorly matched with the system constraints, and results in poor performance. Either a faster controller should be selected, or static checkpoints should be used.

Random and Static Checkpoints The addition of static checkpoints either through the provably secure or heuristic techniques results in the monitoring of the electrodes directly adjacent to the dispense ports. However, there are more dispense ports than checkpoints allowed, so the system designer is only permitted to choose a subset of the static checkpoints. One choice could be to place static checkpoints only within the dispense ports of the reagents and samples, so as to conserve fluids which may be expensive or difficult to obtain. The attacker is thus obligated to pass through a static checkpoint and be detected with high probability. Due to the unique architecture of this chip, there is no way for the attacker to route the attack in a way that avoids these checkpoints. For this particular arrangement and attack, $Pr(E) = 0.49$ (Fig. 4.13) decreasing to $Pr(E) = 0.47$ (Fig. 4.14) when $c = 1$. Most of the benefit comes from the static checkpoint, and we get much better performance without having to increase the checkpoint capacity of the controller.

4.8.3 TNT Detection

Trinitrotoluene (TNT) is a common, widely used explosive chemical compound that is also toxic at extremely low concentrations. Detection of TNT has been of much

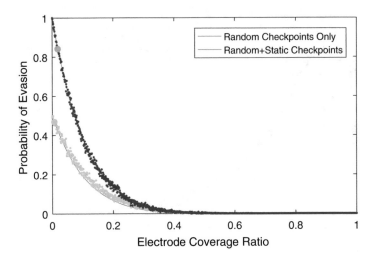

Fig. 4.14 When probability of monitoring random checkpoints increases to 100% ($c = 1$), performance increases only for higher electrode coverage ratios. The low electrode coverage ratio forces $Pr(E)$ to be somewhat high, with static checkpoints showing substantial benefit

interest as part of twenty-first century counterterrorism efforts. To address this need, A DMFB-based platform has been demonstrated for the colorimetric detection of TNT and other nitroaromatic explosives [16].

Detection of TNT on the DMFB is conducted using a colorimetric assay, which is a detection technique commonly employed in microfluidic platforms. Jackson–Meisenheimer colored complexes are formed when nitroaromatic compounds react with nucleophiles. The platform in [16] uses potassium hydroxide (KOH) as the base for detecting the presence of TNT. The absorbance of the sample–reagent mixture exhibits a linear response against the concentration of TNT, as detected by a 505 nm LED and photodiode. Before a sample can be evaluated, a calibration curve must be established, which dilutes known reference samples into several well-defined concentrations. The concentration of the sample-under-test is determined by comparing its absorbance against the calibration curve.

4.8.3.1 Calibration Curve Attacks

The directed acyclic graph for the TNT colorimetric assay is shown in Fig. 4.15. It consists of several chains of dilutions for generating the calibration curve, and one chain for the processing of the sample-under-test. An attacker can alter the final TNT concentration reading by diluting the sample-under-test on the DMFB. One method that can be used to achieve this is to transport one of the waste droplets from dilution chain 3 into the path of the sample droplet in chain 4 [17]. The waste droplet is normally sent to a waste output port for disposal, and is not on a critical path so it is not monitored by error recovery systems. Therefore, this attack can be stealthy.

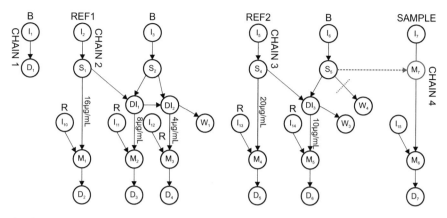

Fig. 4.15 Colorimetric assay for detection of TNT. Each chain dilutes a known reference sample in order to form a calibration curve. The sample is mixed with reagent, and compared against the calibration curve to determine the concentration of TNT. An attacker can dilute the sample by redirecting a waste droplet that is generated in the course of the calibration (shown in dashed lines). B is buffer, REF1 and REF2 are two droplets of known concentration, R is reagent. I_i is dispense, D_i is detect, DI_i is dilute, M_i is mix, S_i is split, and W_i is output waste

Alternately, an attacker could tamper with the assay used for calibration of the readings. The sample being analyzed would then be handled correctly, but interpreted against an incorrect calibration curve.

While protocols could be put in place to enable forensic analysis of the droplets after detection has been completed (e.g., storing all processed fluids for validation), this defeats the purpose of implementing an advanced microfluidic TNT detector in the first place: early detection and notification. A real-time integrity monitoring system is critical for the TNT detector to serve its intended purpose.

4.8.3.2 Defending Against Calibration Curve Attacks

An attacker can redirect a waste droplet to dilute the sample and alter the final reading. Figure 4.16 shows the execution of the DMFB at a time-step where the sample-under-test is being mixed with reagent within the $MIX1$ module. The attacker modifies the actuation sequence such that the waste droplet is moved to merge with the sample droplet (solid red line). Once the sample is diluted, it may no longer meet the concentration threshold required to trigger an alarm. This malicious route requires only one move, and is trivial to implement on an unsecured DMFB system. However, this droplet was specified to be routed to the waste reservoir, and is now missing. The route in the dot-dashed lines represents the missing route generated from the simulation tool. This route takes 19 time-steps, and the effective attack lifetime is 20 time-steps. Therefore, with $k = 42$ checkpoints, our electrode coverage ratio is 0.147, and using Eq. (4.5) we see that the system has $Pr(E) = 0.041$.

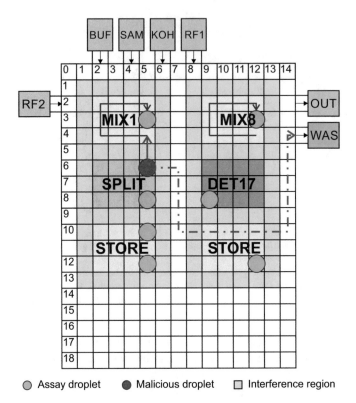

Fig. 4.16 TNT colorimetric assay being executed on a DMFB. The attacker takes the droplet in red and mixes it with the sample in module *MIX*1, diluting it. The route in the dot-dashed line is missing and can be detected. *BUF*, *SAM*, *KOH*, *RF*1, *RF*2 are dispense reservoirs for buffer, sample, potassium hydroxide, reference 1, and reference 2, respectively. *OUT* is the output port and *WAS* is a waste reservoir. Droplets in *MIX*, *SPLIT*, *DET*, *STORE* regions are being mixed, split, detected, and stored, respectively

In other words, there is greater than 95% chance that the randomized checkpoint system will succeed and detect anomalies resulting from the dilution attack. If even higher security is desired, a more powerful coprocessor could be selected to increase k. Conversely, if the application can afford to trade off cost for security, k can be decreased while maintaining a finite probability of success.

4.8.4 Discussion

The probability of evasion achieved in these case studies provides a strong disincentive for would-be attackers. A probability of evasion equal to a fair coin flip is easy to obtain and can be achieved even if the electrode coverage ratio is less than 10% (Fig. 4.10), using both random and static checkpoints being monitored with 50% probability each cycle. Forcing the random checkpoints to be monitored on every cycle causes the probability of evasion curves to drop (Figs. 4.11 and 4.14). The situation only gets worse for the attacker the longer an attack takes place, as the lifetime L gives exponentially lower probability of evasion. Even if we consider other types of attacks where droplets are routed between modules where L is lower, and consequently, $Pr(E)$ is higher, evaluation of probabilistic models likely underestimates the real-world effect of implementing such a system. The fact that a randomized checkpoint system exists is a deterrent to any would-be attackers.

Interpretation of the probability of evasion is not straightforward, and depends on several factors outside the control of the system designer. Certain applications can be expected to have a higher tolerance for risk than others. For instance, assays used for the determination of medical decisions likely have a much lower threshold for probability of evasion than for a lone researcher working on a single experiment. Looking from the perspective of an attacker, the probability of evasion must be sufficiently high since the consequences of being detected may be devastating; a company conducting corporate sabotage needs to be near absolutely certain that an attack cannot be detected. Furthermore, a finite probability of detection would detract from the cost–benefit analysis of developing sophisticated hacking techniques—the Stuxnet worm is believed to have been developed with the resources of a nation-state [18]. If the attackers had to target PLC controllers with security hardware in place, they may have focused on other, perhaps non-technical, means of thwarting Iran's nuclear program. Thus detection systems provide great disincentive for would-be attackers.

4.9   Summary and Conclusion

This chapter presented the analysis and design of an intrusion detection system targeting actuation tampering attacks on a DMFB-based CPMB. We showed that the probability of evasion is largely determined by the length of the malicious route, and that the uniform distribution is secure for general-purpose DMFB arrays. Static checkpoints were introduced in order to influence malicious droplets to take a circuitous path while enforcing critical electrodes, and both provably secure and heuristic placement algorithms were presented. We demonstrated how existing error recovery schemes contribute to the security of the DMFB system. The concept was implemented using an open-source DMFB simulation tool and evaluated on biochips implementing a 3-plex immunoassay, polymerase chain reaction, and TNT detection. The simulation results evaluated the probability of an attack evading detection, and the results supported the design intuition and analysis presented.

The main strength of this approach is its practicality and ease of implementation. There are no subtle security protocols that can be undermined by an incorrect implementation, which often occurs with security technologies such as encryption. The approach is sound due to the principle of physical separation; the network-connected components allow data transmission, while the network-blind components perform auditing. And, by monitoring the physical bioassay progression, we can detect malicious activity that would be missed by hashing the actuation sequence or code signing.

The proposed defense has some limitations. The golden actuation sequence must be furnished to the security coprocessor before deployment, eliminating the possibility of in-field reprogrammability. Repurposing the DMFB would require a secure bootstrap phase. Also, the countermeasure only considers the presence/absence of droplets. Attributes such as volume and concentration could be considered to increase difficulty for an attacker.

While many elaborate ideas for security engineering are being pursued in academic journals and conferences, these technologies are often unproven and may be difficult to implement in practice. In contrast, engineers can start implementing a randomized checkpoint system to defend against attacks today.

References

1. Y. Luo, K. Chakrabarty, T.-Y. Ho, Error recovery in cyberphysical digital microfluidic biochips. IEEE Trans. Comput. Aided Des. Integr. Circuits Syst. **32**(1), 59–72 (2013)
2. P.A. Williams, A.J. Woodward, Cybersecurity vulnerabilities in medical devices: a complex environment and multifaceted problem. Med. Devices Evid. Res. **8**, 305 (2015)
3. Z. Hua, J.L. Rouse, A.E. Eckhardt, V. Srinivasan, V.K. Pamula, W.A. Schell, J.L. Benton, T.G. Mitchell, M.G. Pollack, Multiplexed real-time polymerase chain reaction on a digital microfluidic platform. Anal. Chem. **82**(6), 2310–2316 (2010)
4. V. Srinivasan, A digital microfluidic lab on a chip for clinical diagnostic applications. PhD thesis, Duke University, 2005
5. S. Kennedy, *PCR Troubleshooting and Optimization: The Essential Guide* (Horizon Scientific Press, Poole, 2011)
6. T.W.S. Journal, Theranos Results Could Throw off Medical Decisions, Study Finds (2016)
7. A. Barenghi, L. Breveglieri, I. Koren, D. Naccache, Fault injection attacks on cryptographic devices: theory, practice, and countermeasures. Proc. IEEE **100**(11), 3056–3076 (2012)
8. C.Y. Lee, An algorithm for path connections and its applications. IEEE Trans. Electron. **EC-10**(3), 346–365 (1961)
9. F. Hadlock, A shortest path algorithm for grid graphs. Networks **7**(4), 323–334 (1977)
10. J. Hao, J.B. Orlin, A faster algorithm for finding the minimum cut in a graph, in *Proceedings of the ACM-SIAM Symposium on Discrete Algorithms* (Society for Industrial and Applied Mathematics, Philadelphia, 1992), pp. 165–174
11. M. Stoer, F. Wagner, A simple min cut algorithm, in *European Symposium on Algorithms* (Springer, Berlin, 1994), pp. 141–147
12. D. Grissom, K. O'Neal, B. Preciado, H. Patel, R. Doherty, N. Liao, P. Brisk, A digital microfluidic biochip synthesis framework, in *2012 IEEE/IFIP 20th International Conference on VLSI and System-on-Chip (VLSI-SoC)* (IEEE, Piscataway, 2012), pp. 177–182
13. D. Grissom, P. Brisk, A field-programmable pin-constrained digital microfluidic biochip, in *2013 50th ACM/EDAC/IEEE Design Automation Conference (DAC)* (IEEE, Piscataway, 2013), p. 46
14. R. Higuchi, C. Fockler, G. Dollinger, R. Watson, Kinetic PCR analysis: real-time monitoring of DNA amplification reactions. Biotechnology **11**, 1026–1030 (1993)

15. R. Sista, Z. Hua, P. Thwar, A. Sudarsan, V. Srinivasan, A. Eckhardt, M. Pollack, and V. Pamula, Development of a digital microfluidic platform for point of care testing. Lab. Chip **8**(12), 2091–2104 (2008)

16. V. Pamula, V. Srinivasan, H. Chakrapani, R. Fair, E. Toone, A droplet-based lab-on-a-chip for colorimetric detection of nitroaromatic explosives, in *18th IEEE International Conference on Micro Electro Mechanical Systems* (IEEE, Piscataway, 2005), pp. 722–725

17. S.S. Ali, M. Ibrahim, O. Sinanoglu, K. Chakrabarty, R. Karri, Security assessment of cyberphysical digital microfluidic biochips. IEEE/ACM Trans. Comput. Biol. Bioinform. **13**(3), 445–458 (2016)

18. N. Anderson, Confirmed: US and Israel created Stuxnet, lost control of it (2012)

Chapter 5
Mitigation: Tamper-Mitigating Routing Fabrics

5.1 Introduction

Mitigation techniques aim to provide some assurance of service quality despite the presence of an attack. Mitigation techniques can be active or passive. Active methods leverage an attack detection methodology (such as optical checkpoints) to take decisive action. Examples of active methods include error recovery [1] and feedback stabilized systems. Passive methods lessen the effect of an attack intrinsically through architectural choices or material properties. Examples of passive mitigation techniques in computer architecture include redundancy in time or space.

In this chapter, we study actuation tampering mitigation techniques for CPMBs based on reconfigurable flow-based routing fabrics. Intuitively, a biochip designed for a single function is physically unable to realize an undesired operation. On the other hand, a reconfigurable biochip could be configured in a way that is not only undesirable, but potentially destructive. Thus, microfluidic routing fabrics have security implications. We call a routing fabric designed using the methodology described here *tamper-mitigating*, as it probabilistically reduces the consequences of an attack after it has occurred.

We provide a high-level security assessment of microfluidic biochips utilizing routing fabrics, and analyze their security under actuation tampering attacks. We show that under reasonable assumptions, the permissible states of a routing fabric form a probability distribution. We then provide two methods for determining this distribution: an enumerative approach for bit-limited attacks, and a binary tree representation for random attacks. We then show how to synthesize routing fabrics that satisfy a specified distribution. We call a routing fabric designed in such a way *tamper-mitigating*, as it makes the effects of tampering probabilistically less severe. We then show how the proposed methodology can be used to protect microbiology applications from attack.

© Springer Nature Switzerland AG 2020
J. Tang et al., *Secure and Trustworthy Cyberphysical Microfluidic Biochips*,
https://doi.org/10.1007/978-3-030-18163-5_5

We note that there are two design problems associated with microfluidic routing fabrics, each with implications for performance under actuation tampering attacks. The first is architectural synthesis, which is the problem of constructing the routing fabric with routability guarantees. For mason-brick pattern fabrics, sufficient conditions for routing n inputs to m outputs have been derived [2]. The second is the routing problem, which is the determination of transposer states such that the desired input fluids reach the desired output ports. Solutions have been proposed which leverage graph-theoretic algorithms [2, 3]. We do not consider routing here, although it may be a promising area for research.

5.2   Security Assessment

The structure of a typical cyberphysical flow-based microfluidic biochip platform is illustrated in Fig. 5.1. Samples and reagents are loaded manually onto the biochip, which may contain hardware for manipulation of fluids. Inside the platform, the biochip connects to a routing fabric for interface with additional fluid processing blocks such as heaters and detectors. The routing fabric is used to connect different fluid processing elements and reservoirs, but is not used to process fluids directly. It only serves as an interconnection network. Pneumatic control lines are connected from the biochip to actuators that are computer-controlled.

We assume that the biochips under consideration utilize a state-of-the-art process with the ability to integrate valve-based control logic [4–6]. Such a biochip would allow the designer to incorporate blocks such as multiplexers, demultiplexers, and even processors such that *a single pneumatic control line* can drive all the

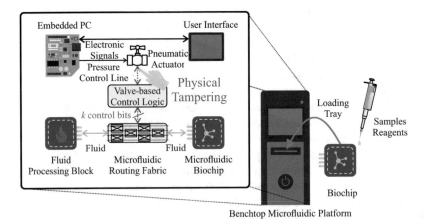

Fig. 5.1 Typical construction of a cyberphysical flow-based microfluidic biochip platform. When deployed as a point-of-care solution, these platforms are prone to physical tampering by end users and can result in undesired operation

control valves in the biochip [7, 8]. Control signals are sent serially over the pneumatic control line and decoded on-chip using valve-based control logic. This is critical as pneumatic control lines are often bulky, with dimensions on the order of millimeters [3]. Recent research [9] has sought to address precisely this issue through pin-constrained design but this is beyond the scope of this work as they are targeted toward more complex biochips.

5.2.1   Threat Model: Physical Tampering

The attacker is a malicious end user or someone with physical access to the microfluidic platform at the point-of-care. They are interested in *stealthily* altering results. Such an attacker poses the greatest threat to undermining the quality of a diagnostic test result as the barrier to tampering is low [10], while motivation is high—spoofed results can be used to falsify compliance when the true result does not match the attacker's desired outcome [11, 12]. Note that this stealthy attack model excludes denial-of-service (DoS) attacks.

Such an attacker, under the tampering taxonomy in [13], is a Class I attacker: a clever outsider who lacks detailed knowledge of the system and has little to no access to sophisticated equipment. Such an attacker will often attempt to leverage existing weaknesses rather than create new ones. We assume that the majority of end users are potential Class I attackers. Expending effort to reduce Class I attacks is warranted given that no evidence exists for Class II (knowledgeable insider) and Class III (funded organization) attacks in the context of microfluidic platforms and medical devices in general.

The attacker induces faulty operation of the routing fabric such that the fluids to be routed are misdirected to the wrong ports. Inducing faults using the fluid control valves leverages the physical vulnerability of the microfluidic platform and requires less expertise than software attacks. Related attacks in the literature are classified as *fault injection attacks*. In cryptographic hardware, such attacks [14, 15] can facilitate cryptanalysis techniques such as differential fault analysis [16].

We assume the attacker is able to open the microfluidic platform to expose the pneumatic control lines. This is reasonable since these platforms are often constructed using sheet-metal chassis and standard screws. Constructing a platform that is physically tamper-resistant would drive up manufacturing costs while reducing serviceability. Further, we assume the attacker does not possess a device that can synchronize injected faults with the biochip controller. Since a single pneumatic control line drives the entire biochip, the attacker has to perfectly time their fault injection attacks in order to precisely control the state of the biochip. The attacker is motivated to tamper with the control state rather than directly swapping the fluid input ports for several reasons:

- Swapping the inputs will likely result in a DoS attack.
- Sensors in the microfluidic platform may easily detect if the samples/reagents at the input reservoirs are incorrect.

- Switching fluids during the execution of an assay through manipulation of valves is stealthy and minimally invasive.
- Routing fabrics can be used as an intermediary between other functional blocks and as such may not have a direct relationship with the inputs of the system.

When the control signals are tampered with, the result is modeled as a change in the control vector from a known state to a randomly selected control vector. The set of control vectors maps into routing fabric states, and therefore, the routing fabric states induce a probability distribution. We note that by considering this threat model, we eliminate "low-hanging fruit" for attackers. Consider the alternative: a biochip with parallel control lines fully exposed. By visual inspection [17], an attacker can precisely change the biochip state.

5.2.2 Attack Implications

Under the previously described threat model, the practical implications of an attack are as follows:

1. *Fluid redirection*: The purpose of a microfluidic routing fabric is to direct a set of fluids from the input ports to the output ports. Under an attack, some fluids may be redirected to the incorrect port. In Fig. 5.2, we see that attacking a single transposer in the routing fabric in dashed lines causes fluids at output ports 1 and 2 to be swapped. In an application where each of the fluids is used for a

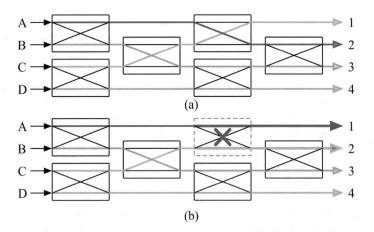

Fig. 5.2 Fluid redirection under attack. (**a**) Each color represents a different fluid flow under normal operation. (**b**) The transposer in the dashed outline is under attack, causing fluids intended for output ports 1 and 2 to swap places

chemical reaction, the fluids at port A and B may be so different as to cause complete failure of the reaction. In a droplet barcoding application, this attack can cause cells to be mislabeled which has consequences for the integrity of scientific inquiry.

2. *Fluid mixing*: If the control signals of a transposer are fully accessible, then it is possible to place the valves into a state where the input fluids mix. Such an attack has consequences that have yet to be fully understood. Since the control valves in a single transposer cannot be actuated simultaneously, fluid mixing can only occur at the architectural level.

3. *Inducement of failure modes*: Reconfiguring the routing fabric into a prohibited state may cause premature failure. Certain hardware primitives may allow multiple inlet valves to flow into the same port, causing excess pressure to build up. Additionally, repeated actuation of the control valves may lead to premature wear. Techniques have been developed to increase the reliability of routing fabrics under disconnection fault models due to valve failure [18].

It is clear that microfluidic routing fabrics are vulnerable to many security threats with serious real-world implications. The problem is thus how to analyze and design routing fabrics with quantifiable security guarantees.

5.3 Problem Overview

Without the ability to synchronize and perfectly alter the serial pneumatic control signal, the attacker is essentially limited to guessing a random control vector. The physical routing that results is therefore randomly chosen from a sample space consisting of all the possible transposer states.

Several transposer states may map into equivalent physical routings. Therefore, some physical routings are more likely than others. The routing fabric architecture induces a probability distribution, and this distribution determines the performance of the fabric under attack (Fig. 5.3). Put another way, given a routing fabric, we would like to know *what will most likely occur if the operator of the microfluidic platform is to lose control of the system?* Then, we would like to know *how do we design a microfluidic routing fabric that mitigates actuation tampering attacks?* These are the analysis and design problems described in Sects. 5.4 and 5.6, respectively.

5.4 Routing Fabric Analysis

In this section, we develop an efficient analysis framework for evaluating security properties of routing fabrics.

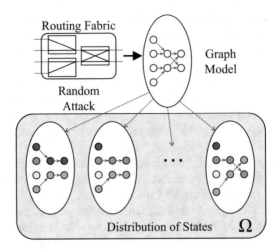

Fig. 5.3 Overview of the problem. A routing fabric admits several states as a function of a control vector. An attacker can only randomly choose control vectors, so the state space forms a probability distribution. We seek to analyze this distribution, and understand how to synthesize routing fabrics with desirable distributions

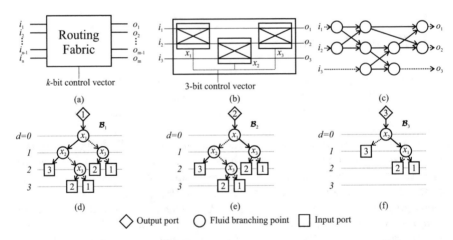

Fig. 5.4 Routing fabric analysis. (**a**) Generic routing fabric with n inputs and m outputs, consisting of k transposers. (**b**) A 3-to-3 routing fabric ($\mathcal{F}_{3 \times 3}$) composed of three transposers. (**c**) Physical graph model used for routing. (**d**) Routing graph models explicitly show how transposer states lead to fluid routings. $\mathcal{B}_1$ has five possible routings. (**e**) Sub-routing graph model $\mathcal{B}_2$. (**f**) Sub-routing graph model $\mathcal{B}_3$. Note that the arrows are opposite to the direction of fluid flow

5.4.1 Modeling Preliminaries

Figure 5.4a illustrates a generic routing block with a set of n input fluids $\{i_1, i_2, \ldots, i_n\}$ and m output ports $\{o_1, o_2, \ldots, o_m\}$. The control port accepts a control vector $s \in \{1, 0\}^k$, where k is the number of binary reconfigurable primitives in the fabric. The result of applying a particular control vector can be observed at the output as a vector with m entries, taking on values from the set $\{1, 2, \ldots, n\}$ to indicate which input fluid is present. If we assume that m, n, k are fixed parameters, then the routing fabric can be interpreted as a function $f : \{1, 0\}^k \rightarrow \{0, 1, 2, \ldots, n\}^m$, where the domain is the set of all control vectors and the range is the set of all output vectors.

We model attacks using a set of random variables $\{X_1, X_2, \ldots, X_k\}$ where each $X_i \sim Bernoulli(0.5)$ corresponds to a transposer control line and k is the number of control bits. This model is based on two assumptions: first, that an attacker randomly perturbs the pneumatic control line as a function of time and second, that the attacker must guess how many faults to inject. In order to know how many faults to inject, an attacker must know the current routing fabric state. Extracting this information is impractical given the physical tampering point-of-care threat model and the fact that multiple routing fabric states can be used to achieve the same fluid routing (i.e., the routing problem [2]). We can define a related random variable $Y_i = f_i(X_1, X_2, \ldots, X_k)$ for each output port o_i that indicates which input fluid i_i appears. The function f_i describes how the routing fabric architecture behaves given a realization of the control vector.

5.4.2 Physical Graph Model

Any routing fabric can be described in terms of an equivalent directed acyclic graph (DAG) $\mathcal{F}_{n \times m} = (\mathcal{D}_{n \times m}, \mathcal{S}_{n \times m})$, where each vertex $d_i \in \mathcal{D}_{n \times m}$ represents a decision point and each edge $s_i \in \mathcal{S}_{n \times m}$ represents a fluid flow channel between the decision points [2, 3]. We call such a representation the *physical graph model* of the routing fabric, as the model can be uniquely mapped to a hardware implementation. This representation can be extended to be state-dependent in order to explicitly model the changes in topology that occur when fluids are actively routed [19, 20]. Consider the 3-to-3 routing fabric in Fig. 5.4b. Its physical graph model is shown in Fig. 5.4c, and shows all the possible fluid pathways. Edges in dashed lines are included for clarity, but are not part of the model used for routing [2].

5.4.3 Routing Graph Model

While directed acyclic graphs provide a physical interpretation of a routing fabric, they obscure the relationship between control and the intended result. Previous studies investigated the effect of different control states using the state-dependent graph [19] representation, where each related subgraph must be generated and evaluated [21]. This is computationally expensive and does not facilitate the design of new routing fabric architectures. To address this problem, we propose to transform the physical graph model into an equivalent graph which explicitly represents the fluid routing possibilities.

Definition 5.1 A routing graph $\hat{\mathcal{B}} = (\hat{\mathcal{V}}, \hat{\mathcal{E}})$ consists of a set of m rooted directed binary trees $\{\mathcal{B}_1, \mathcal{B}_1, \ldots, \mathcal{B}_m\}$, where each $\mathcal{B}_i = (\mathcal{V}_i, \mathcal{E}_i)$ and

1. Each $\mathcal{B}_i$ for $1 \leq i \leq m$ corresponds to fluid output o_m, and is called a sub-routing graph.
2. Each $v_i \in \mathcal{V}_j$ for $1 \leq j \leq m$ represents a transposer-based branching point in the routing fabric reachable by output o_m.
3. Each $e_i \in \mathcal{E}_j$ for $1 \leq j \leq m$ represents a physical pathway between transposers reachable by output o_m.

The root of each sub-routing graph represents one of the output ports. Input ports are represented by leaf vertices (vertices with no outgoing edges) which are labeled to indicate which input port i_n they flow from. Schematically, we will show input vertices as squares and output vertices as diamonds. All other vertices are labeled with x_i, where i corresponds to a transposer and are shown as circles. As we will see, such a representation lends itself to efficient security analysis. The dependencies between states are encoded in the tree structure, while the output states are explicitly represented for intuitive interpretation. Derivation of the routing graph from the physical graph can be done with complexity $O(m \cdot (|V| + |E|))$, using depth-first search or breadth-first search for each of the m outputs. Note that the routing graph model is constructed from the perspective of the output ports and that the directed edges are oriented opposite of fluid flow. This is because of the perspective of the security evaluation in the next section, which is concerned with what input fluid arrives at each output.

Figure 5.4c–f shows the routing graph models of the 3-to-3 routing fabric. Note that each output is interpreted as a separate rooted binary tree. Isomorphic subgraphs could have been shared among the three inputs, forming a multi-rooted shared decision diagram [22, 23]. However, the focus of this work is not finding the most efficient implementation but rather on easily understandable analysis, so we keep the trees separated. Also note that the routing graph model can be derived directly from the routing fabric rather than through the physical graph model. We have described the physical graph model for completeness and to set up a framework for the synthesis phase, where we build a routing fabric using routing graphs as building blocks.

5.4.4   Evaluating Security

Given that an attack is probabilistic under the threat model described previously, we would like to know which input fluid will likely be routed to each output. Therefore, *evaluating the security of a routing fabric is reduced to computing the probability distributions induced by the routing graph model.* Given the routing graph model

Table 5.1 Probability mass functions for the 3-to-3 routing fabric

Variable	$p(y=1)$	$p(y=2)$	$p(y=3)$
Y_1	3/8	3/8	1/4
Y_2	3/8	3/8	1/4
Y_3	1/4	1/4	1/2

$\hat{\mathcal{B}} = \{\mathcal{B}_1, \mathcal{B}_2, \ldots, \mathcal{B}_m\}$ we can calculate the induced probability mass functions (PMFs) $\hat{\mathcal{P}} = \{p(Y_1), p(Y_2), \ldots, p(Y_m)\}$ as follows. Each sub-routing graph has n different leaf vertex labels and therefore each of the m PMFs has n outcome probabilities $\{p_1, p_2, \ldots, p_n\}$. Each of the leaves on the tree has a probability 2^{-d}, where d is the depth of the leaf, since it is assumed that the transposers are i.i.d $Bernoulli(0.5)$. Therefore each of the p_i is found by summing leaf probabilities with the same label. This can be written as a binary expansion:

$$p_i = \sum_{d \in D_i} 2^{-d} \tag{5.1}$$

where D_i is the set of depths associated with leaf vertices labeled i. That is, each probability is constructed out of "atoms" of the form 2^{-d}. This bears similarity to the concept of a *generating tree* by Knuth and Yao [24]. However, generating trees simulate arbitrary distributions by decomposing them into dyadic atoms and assigning them to leaves of the tree (here we only evaluate the distributions). This can be interpreted as a procedure for simulating distributions through the use of a single unbiased coin flip.

Referring back to the examples in Fig. 5.4c–f, we see that each leaf at depth level 2 has probability 1/4, while those at depth 3 have probability 1/8. Summing common leaf vertices, the PMFs are evaluated in Table 5.1. Interpretation of the probability distribution depends on the intent of the designer. We will revisit interpretation and security metrics in Sect. 5.6, after describing applications with interesting desired behaviors. We note that Knuth and Yao's generating tree represents a procedure, or an algorithm, for simulating an arbitrary distribution, while the routing graph in this work is a representation of fluid flow paths. The practical implication of this is that generating trees can be more expressive, as they can have feedback and be infinite. This concept has been extended to show that arbitrary rational distributions can be generated in stochastic flow networks [25].

5.5 Routing Fabric Synthesis

In routing fabric analysis, we start with a complete architecture and decompose it into its constituent parts to evaluate the induced probability distribution. In routing fabric synthesis, we proceed in the opposite direction. We start with a routing graph model and assign transposers to each vertex such that it represents a unique

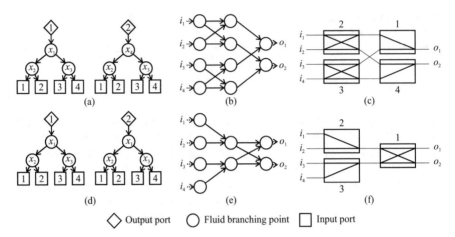

◇ Output port   ◯ Fluid branching point   ☐ Input port

Fig. 5.5 Routing fabric synthesis. (**a**) Non-optimal coloring of vertices for two routing graphs. (**b**) Physical graph model of the non-optimal design. (**c**) Schematic representation of the synthesized routing fabric. Four control lines are required to drive this design. (**d**) Optimal coloring. (**e**) Physical graph model of the optimal design. (**f**) Schematic representation of synthesized optimal routing fabric. Only three control lines are required. However, this design has less routing flexibility than (**c**). Saving even a single transposer has practical benefits, as pneumatic control lines add significant bulk and expense to microfluidic systems

architecture. Constructions of routing graphs will be provided in Sect. 5.6. For now, we assume they are given. In Fig. 5.5a, c we have two alternate transposer assignments for the same routing graph models. In Fig. 5.5a, the root vertices have been assigned different transposers, while in Fig. 5.5b, they are the same. While this is a small difference, it leads to considerable changes in the resultant architecture. Figure 5.5b, c illustrates a synthesized design that uses four transposers, while Fig. 5.5e, f uses only three. The routability of both designs are also drastically different, as Fig. 5.5c allows any input–output permutation, while Fig. 5.5f cannot simultaneously route i_1 with i_2 nor i_3 with i_4.

Transposer assignment can be represented as a vertex coloring problem. By merging pairs of vertices together, we take advantage of the transposer primitive's ability to route two fluids for one control input. We seek to reduce control complexity by minimizing the number of transposers, while ensuring that the resulting design is physically meaningful.

5.5.1   Problem Statement

The synthesis problem statement is defined as follows:

Input: A routing graph model $\hat{\mathcal{B}} = (\hat{\mathcal{V}}, \hat{\mathcal{E}})$.
Output: A synthesized routing graph $(\hat{\mathcal{B}}, c(v))$, where $c(v)$ is a vertex coloring function $v_i \in \hat{\mathcal{V}} \to \mathbb{Z}^+$.

Objective: Minimize the number of transposers.
Constraints: Ensure the design is physically realizable.

5.5.2   ILP-Based Synthesis

We propose to solve the transposer assignment problem using integer linear programming (ILP). Such a formulation naturally permits the description of synthesizability constraints in a graph coloring optimization problem. We will use the term color and transposer interchangeably. We define our model as follows. We take a routing graph as input and divide the vertices into three sets: V_L is the set of leaf vertices, V_R is the set of root vertices, and V_I is the set of all other vertices. That is, $V = V_R \cup V_I \cup V_L$. We denote the number of vertices as $N = |V|$, and the maximum number of colors as $K = |V_I|$.

Let $x_{n,k}$ be a binary variable that equals 1 if vertex v_n has been assigned color k, for $1 \leq n \leq N$ and $1 \leq k \leq K$. Let c_k be a binary variable that is equal to 1 if color k has been assigned to at least one vertex for $1 \leq k \leq K$. This can be modeled as a logical OR of the $x_{n,k}$ color assignment variable, across all vertices:

$$c_k = \bigvee_{1 \leq n \leq N} x_{n,k}, \quad 1 \leq k \leq K \tag{5.2}$$

The optimization goal is to reduce the number of transposers, which means assigning 1 to as few of the c_k's as possible. Therefore, we state the objective function as:

$$\min : \sum_{1 \leq k \leq K} c_k \tag{5.3}$$

Subject to the following constraints:

1. *Limit on transposer input ports:* Each transposer has two distinct inputs and as such can only accept two unique fluid flow paths. This means the set of predecessor colors for each transposer color cannot exceed two. Define an additional binary variable $w_{i,j}$ which equals 1 if color i has color j as a predecessor.

$$\sum_{1 \leq j \leq K} w_{i,j} \leq 2, \quad 1 \leq i \leq K \tag{5.4}$$

2. *Limit on transposer output branching paths:* A transposer can toggle an input fluid between two output ports. Both input ports must obey the same restriction. Therefore each color cannot have more than two different colors as children.

$$\sum_{1 \leq i \leq K} w_{i,j} \leq 2, \quad 1 \leq j \leq K \tag{5.5}$$

Table 5.2 Notation used in ILP formulation

Variable	Description
V_R	Set of root vertices
V_I	Set of intermediate vertices
V_L	Set of leaf vertices
K	Number of possible color assignments
N	Number of vertices in the routing graph
$x_{n,k}$	0–1 variable indicating if vertex v_n is assigned color k
c_k	0–1 variable indicating if color k has been used
$w_{i,j}$	0–1 variable indicating if color i has j as predecessor

3. *Unintentional mixing:* Leaf vertices must be driven by only a single transposer. To do otherwise would imply that two independently controlled paths are connected to the same output port, meaning there exists a configuration of transposers that would result in fluids mixing.

$$\sum_{1 \le j \le K} w_{i,j} \le 1, \quad \forall i \in V_L \tag{5.6}$$

This constraint can be removed to include mixing states.

4. *All paths must have an assignment:* Each vertex in the routing graph must be assigned exactly one transposer.

$$\sum_{1 \le k \le K} x_{n,k} = 1, \quad 1 \le n \le N \tag{5.7}$$

We summarize the notation used in our model in Table 5.2.

5.5.3 Fast Synthesis

The ILP model is optimal; however, the use of a general-purpose ILP solver may require unacceptable computational time as they explore a search space without exploiting any domain-specific knowledge. Here, we show that the special restricted variant of the synthesis problem as described in the previous section is solvable in polynomial time. The ILP formulation is still useful for precisely specifying the problem, checking for correctness, and accommodating alternate constraints that may or may not yield a polynomial-time solution.

We propose a fast synthesis algorithm that proceeds by sequentially evaluating unassigned vertices and deciding whether to add them to the current transposer or start a new transposer (Fig. 5.6). At the beginning of the routine, we form a new transposer and iterate through unassigned vertices to find a suitable candidate

Input: Routing graph $\hat{\mathcal{B}} = (\hat{\mathcal{V}}, \hat{\mathcal{E}})$
Output: Map $c(v)$ from vertices to colors
1: $c(v) \leftarrow \varnothing, \forall v \in \hat{\mathcal{V}}$
2: $currColor \leftarrow 1$
3: **while** $\exists \varnothing \in c(v)$ **or** new assignments made **do**
4:     **for each** $\{v_i \in \hat{\mathcal{V}} : c(v) = \varnothing\}$ **do**
5:         **if** v_i's children have assigned color **then**
6:             $c(v_i) \leftarrow currColor$
7:             break
8:         **end if**
9:     **end for**
10:     **for each** $\{v_i \in \hat{\mathcal{V}} : c(v) = \varnothing\}$ **do**
11:         **if** v_i's children colors match currColor's children colors **then**
12:             **if** $currColor$ already has two parent colors **then**
13:                 $currColor \leftarrow currColor + 1$
14:                 $c(v_i) \leftarrow currColor$
15:                 break
16:             **end if**
17:             $c(v_i) \leftarrow currColor$
18:         **end if**
19:     **end for**
20:     $currColor \leftarrow currColor + 1$
21: **end while**
22: **if** $\{\exists \varnothing \in c(v)\}$ **then**
23:     **throw** $Infeasible Solution Exception$
24: **end if**
25: **return** $c(v)$

Fig. 5.6 Fast routing graph synthesis pseudocode

for linking. Linking is performed if a vertex's two children already have a color assignment. At startup, the only vertices that satisfy this property are leaf vertices. After the first assignment, we iterate through the remaining unassigned vertices and see if they can be shared. Sharing is performed if a vertex's child vertices have the same colors as the ones already associated with the transposer. Once all vertices have been enumerated, a new transposer is formed. This repeats until all vertices are assigned. Note that an infeasible solution occurs when no new assignments have been made between iterations of the loop and there exist vertices without a color assignment.

Theorem 5.1 *The fast synthesis algorithm solves the transposer assignment problem optimally in $O(N^2)$.*

Proof We prove optimality in two parts.

(i) Assume a routing graph consisting of complete binary trees with colored leaf vertices. Consider any arbitrary set $\mathcal{D}$ of vertices such that the children of these vertices have a color assignment. Let m be the number of colors assigned to

$d \in \mathcal{D}$. m is minimized when all vertices $d \in \mathcal{D}$ with the same set of children colors are assigned the same color. This shows how to minimally color a set of vertices.

(ii) Let n be the number of colors assigned to the children of $\mathcal{D}$, and assume that initially $n = n_0$ and $m = m_0$. If we change the color assignments of the child vertices such that $n > n_0$, then $m \geq m_0$. That is, the number of colors used in $\mathcal{D}$ cannot decrease as introducing new child colors will no longer permit sharing of colors in $\mathcal{D}$. For any set of vertices, the number of colors is optimized when the number of children colors are minimized.

Therefore, assigning colors according to (i) minimizes the colors for a given set of vertices, and also sets up the minimizing conditions (ii) for the parent vertices of this set.

For worst-case complexity, we observe that the algorithm consists of one outer loop and two inner loops. The first inner loop iterates through all unassigned vertices until a suitable candidate is found for adding to the current color, which is $O(|\hat{\mathcal{V}}|)$. The second inner loop iterates through all unassigned vertices, excluding the one that was just assigned, which is $O(|\hat{\mathcal{V}}|)$. At each iteration of the outer loop, the worst-case behavior is that only a single color is assigned to each vertex, so this completes in $O(|\hat{\mathcal{V}}|)$ as well. Therefore, the overall worst-case complexity is $O(|\hat{\mathcal{V}}|^2)$.

Theorem 5.2 *The routing fabric synthesis problem is polynomial-time solvable in* $O(|\hat{\mathcal{V}}|^2)$.

Proof This follows directly from the preceding theorem. $\square$

5.5.4 Routing Graph Reduction

After the routing graph has been synthesized, we can convert it into a physical graph model using a reduction algorithm, similar to the reduce process used in binary decision diagrams [26]. Leaf vertices map to unique output ports, so all leaf vertices with the same label must be merged together, with all incoming edges redirected to the merged vertex. Intermediate vertices with the same label and with matching predecessors can be merged together. Note that vertex labels can appear twice in the final physical graph model, as they represent unique input ports on a single transposer. The common label refers to the shared control. This can be performed quickly using the algorithm in Fig. 5.7 with complexity $O(|\hat{\mathcal{V}}| + |\hat{\mathcal{E}}|)$. The unique colors in the map $c(v)$ define vertices in the physical model. We then enumerate the edges in the routing graph and translate them into edges in the physical model based on the color of the edge's vertices. Then we delete redundant edges, and expand vertices with two incoming edges. The transformed graph is a physical graph model. This physical graph model represents a completed routing fabric design with desirable security properties when driven by a serial control line.

Fig. 5.7 Routing graph
reduction pseudocode

Input: Routing graph $\hat{\mathcal{B}} = (\hat{\mathcal{V}}, \hat{\mathcal{E}})$, map $c(v)$
Output: Physical graph $\mathcal{F}_{n \times m}$
1: **for each** $\{e = \{s, t\} \in \mathcal{E}\}$ **do**
2:      $s \leftarrow c(s), t \leftarrow c(s)$
3: **end for**
4: delete redundant edges in $\mathcal{E}$
5: **for each** $\{v \in \mathcal{V}$ with two incoming edges$\}$ **do**
6:      create duplicate of vertex $\hat{v} = v$
7:      redirect one of the incoming edges of v to $\hat{v}$
8: **end for**
9: **return** $\mathcal{F}_{n \times m}$ = modified $\hat{\mathcal{B}}$

5.5.5   Caveats

Note that only the specific instance of transposer assignment as described in the ILP model is polynomial-time solvable. Changing the ILP model to accommodate alternate primitives or to permit mixing of fluids may not yield fast solutions. It is an open question as to what effect the ordering of leaf vertices has on optimality of transposer assignment. Furthermore, the synthesis methodology presented here does not specify how the input routing graphs are generated. The routing graphs can satisfy any functionality (within some practical limitations) desired by the system designer, as we will demonstrate in the following design example.

5.6   Application: Forensic DNA Barcoding

We now turn to the development of some simple constructions to generate routing fabrics with desirable security properties for a forensic DNA barcoding application, which is ripe for tampering at the point-of-care [27]. Microfluidic technology has made high-throughput DNA analysis and barcoding a reality, allowing scientists to study gene expression as a function of cell type. Cellular analysis is a widely used procedure in clinical diagnostics, pharmaceutical research, and forensic science [28]. With evidence that cells, even within the same clonal population, are heterogeneous in their genomic responses, a large number of single-cell analysis methods have been established using microfluidic devices [29]. Single-cell analysis relies on encapsulation of individual cells inside droplets and tagging these droplets with unique DNA barcodes; this procedure is referred to as DNA barcoding [30]. Barcoded samples can then be processed through a variable sequence of biochemical operations, while their genomic identity is preserved. To control the DNA barcoding of thousands of heterogeneous cells, a microfluidic routing fabric has been efficiently used [2].

5.6.1 Security Implications

The routing fabric utilized in [2] has six levels, and permutes eight inputs to two outputs (Fig. 5.8a, b). Eight types of barcoding droplets, identified by the letters A–H, are connected to the input ports, to be dispensed on-demand to any of the two output ports. Once dispensed, these droplets are routed to the rest of the platform for further processing and sensing using digital microfluidic technology integrated with actuators and optical detectors.

To illustrate what can potentially go wrong, we show an example fluid routing in Fig. 5.8a. Barcode type A is to be dispensed to port 2, while barcode type F is to be dispensed to port 1. These barcodes are intended for simultaneous application to two cell droplets. If an attacker causes a fault through the serial control line such that all transposer states happen to be swapped, we see that barcode A is halted at an intermediate transposer. Barcode F is still able to make it to the correct port. Barcode H, which was never intended for use at this point in the protocol, is now dispensed to port 2.

The practical implication of this attack for the cell study is that the biochemical procedure will provide misleading outcomes. As a consequence of this attack, a

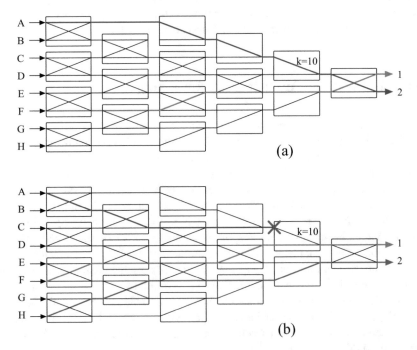

Fig. 5.8 (**a**) Routed fabric used in a DNA barcoding application under normal operation. Fluid A (blue) is routed to output port 2, while fluid F (green) is routed to output port 1. (**b**) After all transposers are attacked, fluid H (red) ends up at output port 2, while fluid F (green) moves to port 1. Fluid A (blue) is blocked at an intermediate transposer

cell that has been identified as type A (based on the *in vivo* activity of a certain biomarker) will be wrongly tagged with a barcode that belongs to a different sub-population of type H. During the process of biomolecular analysis, cells are lysed and type-driven DNA analysis is applied by using the barcode. Hence, the alteration of barcoding causes the gene reads of type-A cells to be interpreted as a part of type-H genomic landscape, thus leading to a false conclusion on the gene expression of type-A cells. If this routine analysis is carried out as a part of a DNA forensic investigation, a suspect (with type-A cells collected from a crime scene) may eventually escape prosecution.

Single-cell applications such as DNA forensics assume that the cells under study were collected and barcoded in a trustworthy manner and therefore allow making clinical or judicial decisions based on the genomic study. As we can see, an attacker could potentially skew the barcoding process by launching attacks on the control vectors for routing fabrics. To secure the barcoding platform against such attacks, we can use the methodology presented in this work. First, ensure that the mapping between control vectors and control valves cannot be determined by the attacker, and then design the routing fabric in such a way that random perturbations cannot produce a meaningful result.

5.6.2   Tamper-Mitigating Routing Graph Constructions

We demonstrate how our synthesis methodology can be used with routing graphs designed from scratch to secure against actuation tampering attacks. The simple constructions developed here can be adapted for other uses, but more interestingly, the general formulation means that new constructions can be tailored to the application.

In the DNA barcoding application, we would like to prevent the attacker from biasing the barcoding process using the concept of *uncontrollability* as follows (Fig. 5.9a):

Definition 5.2   A routing fabric is **uncontrollable** if an attacker is unable to reliably route a specific input fluid to specific output port.

In other words, the PMF $p(Y_i)$ is a uniform distribution. We extend the definition of uniform distribution to mean that the PMF will either take on uniform values or 0. This is to accommodate fabric designs where certain outcomes are not possible.

Security Metric: Normalized Entropy   The information-theoretic concept of entropy naturally captures the security of the distribution:

$$\hat{H}(Y_n) = - \sum_{y_i \in S} \frac{p(y_i) \log_2(p(y_i))}{\log_2(|S|)} \tag{5.8}$$

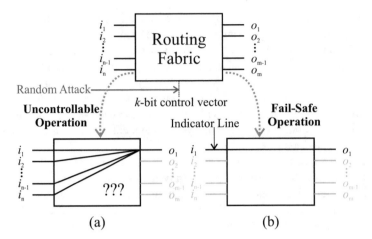

Fig. 5.9 (**a**) Uncontrollable operation. Under attack, it is unknown which input fluid will be directed to a particular output. (**b**) Fail-safe operation. One routing is more likely than all others, and can be connected to an "indicator line" where some inert low-cost fluid is easily detected at the output

where S is the set of all desired outcomes and the $\log_2$ term expresses the entropy in units of bits. By "desired outcomes" we mean to exclude inputs that we may wish to behave in a biased manner, as we will see when we introduce the concept of indicator lines. It is well-known that for discrete distributions, entropy is maximized when the distribution is uniform [31]. We use the normalized form of entropy so that the quantity can be compared across routing fabrics irrespective of the number of input ports n.

Construction If we assume that n is a power of 2 (i.e., the number of input ports is dyadic), then a balanced binary tree of depth $\log(n)$ implements a uniform distribution with the minimum number of vertices. However, it may be desirable to realize a tree of greater depth in case the synthesis algorithm returns an infeasible solution. We define a parameter r as the redundancy of an uncontrollable routing graph, which increases the number of branching points while keeping the induced probability distribution constant. We construct an r-redundant uncontrollable routing graph by duplicating the $r-1$ routing graph and connecting the two subgraphs with one branching point (Fig. 5.10).

As a secondary countermeasure, we would also like to detect if an attack has occurred. This can be achieved by using an indicator fluid. In normal operation, the microfluidic platform will never route the indicator fluid to any of the output ports. Under attack, the routing fabric will dispense indicator fluid with high probability, which can easily be detected by on-board sensors. An indicator fluid can be as simple as water with dye. Detection can be implemented with a *fail-safe* routing fabric (Fig. 5.9b).

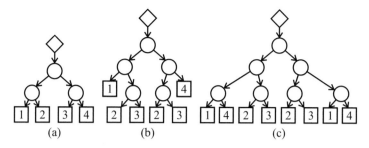

Fig. 5.10 (**a**) Optimal balanced tree construction for $n = 4$ outcomes. (**b**) r-1 expansion of the optimal tree. (**c**) r-2 expansion of the optimal tree. Expansions can continue indefinitely by selecting two leaf vertices of the same depth and converting them into branching vertices with two children with the same labels

Definition 5.3 A routing fabric is **fail-safe** if, under a random control vector, one fluid routing is more likely than all others.

Security Metric: Relative Likelihood We introduce factor ϵ_n as the ratio between the most probable outcome and the second most probable outcome corresponding to Y_n. Let p_m be the m-th highest probability defined by the probability mass function (PMF) $p(Y_n)$. Then

$$\epsilon_n = \frac{p_1(Y_n)}{p_2(Y_n)} \tag{5.9}$$

A PMF with a uniform distribution will thus have $\epsilon = 1$, while on the other extreme, a $Bernoulli(1)$ distribution will have $\epsilon = \infty$. It is desirable to maximize this quantity.

Construction The requirement that one outcome be much more likely than all other outcomes places little restriction on the distribution of the unlikely outcomes. Therefore, instead of attempting to provide an optimal construction, we simply provide a simple method for augmenting an existing routing graph to become fail-safe: define a new root vertex, and connect the old root vertex to one of the outgoing edges, and connect the fail-safe outcome to the other outgoing edge. The relative likelihood is thus $\epsilon \geq 2$, since all of the original outcome probabilities are now bounded by $1/2$ (Fig. 5.11). Intuitively, connecting an extra transposer in this manner extends the state space such that when this extra transposer is active, the routing fabric functions normally, and when it is inactive, the routing fabric is diverted to an indicator line.

We then propose an iterative design procedure, where we start with a minimal construction and then perform r-expansion until a feasible solution is found (Fig. 5.12).

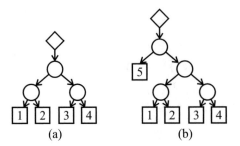

Fig. 5.11 (**a**) Original routing graph taken as input. (**b**) Augmented routing graph where a new fail-safe state 5 has been introduced. Root-level augmentation guarantees that $\epsilon \geq 2$

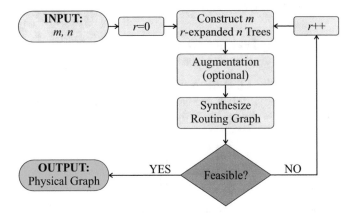

Fig. 5.12 A simple iterative procedure to generate n-to-m routing fabrics with uniform distributions and optionally, a fail-safe mode. Start with the minimal construction achieving the design goals, and incrementally r-expand the routing graphs until a feasible solution is found

5.6.3 Experimental Results

Using the constructions derived here, we generated two identical sub-routing graphs. First, we generate an uncontrollable routing graph to prevent biasing the droplet barcodes toward any particular label. Then we apply the augmentation operation to provide fail-safe functionality. The construction is shown in Fig. 5.13a.

The result of applying either the ILP-based or heuristic synthesis algorithms on these routing graph constructions is shown in Fig. 5.13b. By design, the probability distribution induced by this routing fabric is well-defined. The only downside to this design is that pipelining operations are not defined. The induced probability mass functions are shown in Table 5.3. We see that this work is both fail-safe and uncontrollable, while the original routing fabric is ambiguous as to what guarantees it can provide. We evaluate relative likelihood $\epsilon = 8$ and $\epsilon = 5.33$ for this work and the augmented CoSyn fabric, respectively. Ignoring the fail-safe states induced

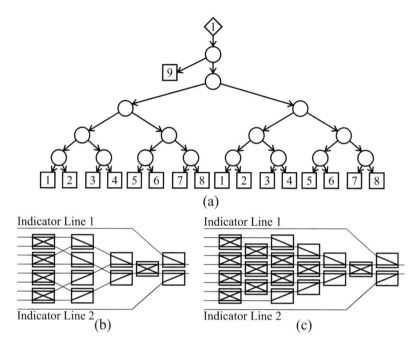

Fig. 5.13 (**a**) Routing graph construction. Two identical graphs are used as input to the synthesis routine. (**b**) Result from synthesis. The tamper-mitigating routing fabric achieves a controlled state distribution, with the same routing flexibility, while using fewer transposers. (**c**) Original 8-to-2 routing fabric design as described in [32]

	Augmented CoSyn [2]		This work	
n	$p(Y_1 = n)$	$p(Y_2 = n)$	$p(Y_1 = n)$	$p(Y_2 = n)$
1	0.1875	0.1875	0.0625	0.0625
2	0.1875	0.1875	0.0625	0.0625
3	0.1563	0.1563	0.0625	0.0625
4	0.1563	0.1563	0.0625	0.0625
5	0.0938	0.0938	0.0625	0.0625
6	0.0938	0.0938	0.0625	0.0625
7	0.0625	0.0625	0.0625	0.0625
8	0.0625	0.0625	0.0625	0.0625
9	0.5000	0.0000	0.5000	0.0000
10	0.0000	0.5000	0.0000	0.5000
$\hat{H}(Y_n)$	0.48	0.48	1.00	1.00
ϵ_n	5.33	5.33	8.00	8.00

Table 5.3 Probability mass functions and security metrics for 8-to-2 routing fabrics

by the indicator lines, we evaluate normalized entropy as $\hat{H} = 1$ and $\hat{H} = 0.48$, showing that the synthesized fabric maximizes uncertainty for the attacker.

What this means for an attacker is that with probability equal to a fair coin flip, any attempt to control the routing fabric will result in their attack being detected. If they happen to succeed in activating the fail-safe mode, then they can only select a droplet barcode with uniform probability. That is, they will be unable to bias the outcome through repeated attacks. Furthermore, repeated attacks will exponentially increase the likelihood of inducing the failure more. In comparison with the original 8-to-2 routing fabric, the attacker would be able to bias the barcoding operation toward barcodes located at ports 1 and 2. This will introduce false biases into the cell study, leading to incorrect results.

5.7   Relation to Prior Work

This work is the first to address both the analysis and synthesis of secure microfluidic routing fabrics. However, our concepts share similarities with many prior works, which we mention here for the interested reader.

- *Graphical models*: Algebraic decision diagrams [23] and their precursors, ordered binary decision diagrams [26], have lent themselves to efficient manipulation of Boolean functions and as such have had a tremendous impact on the CAD industry since their introduction in the late 1970s. This work does not leverage features that have become synonymous with BDDs, such as canonicity through graph reduction and ordering of variables.
- *Probabilistic graphical models*: The physical graph model could be interpreted as a Bayesian network, and the act of determining most probable outcomes as Bayesian inference [33]. However, the analysis of transposers is slightly more complicated due to the interdependence of fluid routing paths and their control; physical graph models contain repeated vertices.
- *Scattering-based analysis*: Tang et al. [21] analyzed security performance in terms of the probability of scattering, where scattering is defined as a fluid being redirected more than d coordinate spaces away from its desired destination. This was also evaluated under the assumption of a certain attack class, organized by the number of bit flips induced by the attack. Such a model presumed the attacker was limited in attack capability while still being forced to randomly choose control lines to tamper with. Here, we argue that such a limitation is unrealistic in practice and that introducing randomness leads to more feasible analysis and design.
- *Fault-tolerant design*: This work differs from general fault-tolerant architecture research in several important ways [34]. First, there is no known transposer failure mode that manifests itself as a flipped state; doing so would require the simultaneous transient failure of several valves [18]. Only an attacker is capable of inducing such a state. Second, while many fault-tolerant systems incorporate

redundancy that can be leveraged in case of failure, many of these systems require the detection of a fault and active redirection of resources. During an attack, no such advantage may be conferred since the attacker may be actively tampering with these systems.

- *Interconnection networks*: And lastly, we note that the concept of a microfluidic routing fabric has historical precedent in interconnection networks. The study of interconnection networks began with the introduction of the Clos network [35], which was designed for telephony applications. In Clos' original paper, conditions were proved for ensuring that the network is non-blocking. Other interconnection topologies include Banyan and Omega networks [36]. Later, these concepts were adopted for networking and computer architectures, and analyzed for their fault-tolerant behavior. The routing fabrics considered in this work are of a mason-brick topology and are not employed in these related fields.

5.8 Summary and Conclusion

We presented a security assessment of an emerging microfluidic hardware primitive: the transposer-based routing fabric driven through a serial control line. We then formulated an analysis and design methodology under a random control vector attack model. Two classes of security, fail-safe and uncontrollable, were defined and case studies were presented to show how such characteristics could be leveraged in practice as a tamper mitigation mechanism. Such a design-time tamper mitigation technique eliminates one of the simplest methods for tampering with a physically vulnerable device employed at the point-of-care.

One major limitation of the framework in this chapter is the restriction on distributions being dyadic. As shown in recent works, feedback greatly increases the expressivity of flow networks, permitting arbitrary rational distributions in compact form [25]. Without feedback, an infinite number of branching points are required [24]. As feedback is prohibited in microfluidic flow networks, it would be an interesting work to see if arbitrary distributions could be efficiently approximated for implementation in a routing fabric. Other limitations include: the fact that pipelining cannot be specified as a design criterion [2], and the number of transposers in the final design is not guaranteed to be minimized.

If we were to consider DoS attacks, then the performance of the routing fabrics in this work would be degraded. For instance, an attacker could perturb only a single transposer in the hope of causing a failure. The fail-safe mechanism would then be activated with probability $1/K$ instead of $1/2$, where K is the number of transposers. Evaluating the probability that a fluid input would deviate from its correct path must take the current transposer state into account and would thus require scattering-based analysis as presented in [21].

Future work could address the problem of designing a tamper-mitigating routing fabric with security guarantees at the *joint* probability distribution level. This is especially challenging, given that the joint probability distributions we consider are actually functions of probability distributions defined through multiple graphical models. We also note that our work is a *passive* mitigation technique, as it requires no active input or control to work. Alternate mitigation techniques could use active feedback to correct anomalous behavior. New microfluidic routing primitives could be investigated; it was noted that in [3] that multiplexer–demultiplexer pairs could have been implemented instead of the transposer primitive.

Finally, we note that this work is among one of the first hardware architecture-based defenses against tampering in microfluidics. The intent is not to replace well-designed cyberphysical systems implementing standard security measures such as encryption. Rather, this work aims to bolster security in the physical domain, which is often overlooked.

References

1. Y. Luo, K. Chakrabarty, Design of pin-constrained general-purpose digital microfluidic biochips. IEEE Trans. Comput. Aided Des. Integr. Circuits Syst. **32**(9), 1307–1320 (2013)
2. M. Ibrahim, K. Chakrabarty, U. Schlichtmann, CoSyn: efficient single-cell analysis using a hybrid microfluidic platform, in *Design, Automation & Test in Europe Conference & Exhibition (Lausanne)* (2017)
3. R. Silva, S. Bhatia, D. Densmore, A reconfigurable continuous-flow fluidic routing fabric using a modular, scalable primitive. Lab. Chip **16**(14), 2730–2741 (2016)
4. I.E. Araci, P. Brisk, Recent developments in microfluidic large scale integration. Curr. Opin. Biotechnol. **25**, 60–68 (2014)
5. S.-J. Kim, D. Lai, J.Y. Park, R. Yokokawa, S. Takayama, Microfluidic automation using elastomeric valves and droplets: reducing reliance on external controllers. Small **8**(19), 2925–2934 (2012)
6. B. Mosadegh, T. Bersano-Begey, J.Y. Park, M.A. Burns, S. Takayama, Next-generation integrated microfluidic circuits. Lab. Chip **11**(17), 2813–2818 (2011)
7. P.N. Duncan, S. Ahrar, E.E. Hui, Scaling of pneumatic digital logic circuits. Lab. Chip **15**(5), 1360–1365 (2015)
8. M. Rhee, M.A. Burns, Microfluidic pneumatic logic circuits and digital pneumatic micropro-cessors for integrated microfluidic systems. Lab. Chip **9**(21), 3131–3143 (2009)
9. M. Ibrahim, A. Sridhar, K. Chakrabarty, U. Schlichtmann, Sortex: efficient timing-driven syn-thesis of reconfigurable flow-based biochips for scalable single-cell screening, in *Proceedings of IEEE/ACM International Conference on Computer-Aided Design* (2017), pp. 623–630
10. G.J. Kost, Preventing medical errors in point-of-care testing: security, validation, performance, safeguards, and connectivity. Arch. Pathol. Lab. Med. **125**(10), 1307–1315 (2001)
11. R. Garver, C. Seife, FDA let drugs approved on fraudulent research stay on the market (2013). https://www.propublica.org/article/fda-let-drugs-approved-on-fraudulent-research-stay-on-the-market
12. H. Fereidooni, J. Classen, T. Spink, P. Patras, M. Miettinen, A.-R. Sadeghi, M. Hollick, M. Conti, Breaking fitness records without moving: reverse engineering and spoofing Fitbit, in *International Symposium on Research in Attacks, Intrusions, and Defenses* (Springer, Berlin, 2017), pp. 48–69
13. D.G. Abraham, G.M. Dolan, G.P. Double, J.V. Stevens, Transaction security system. IBM Syst. J. **30**(2), 206–229 (1991)

14. A. Barenghi, L. Breveglieri, I. Koren, D. Naccache, Fault injection attacks on cryptographic devices: theory, practice, and countermeasures. Proc. IEEE **100**(11), 3056–3076 (2012)
15. H. Bar-El, H. Choukri, D. Naccache, M. Tunstall, C. Whelan, The sorcerer's apprentice guide to fault attacks. Proc. IEEE **94**(2), 370–382 (2006)
16. E. Biham, A. Shamir, Differential fault analysis of secret key cryptosystems, in *Proceedings of Annual International Cryptology Conference (Santa Barbara, CA)* (Springer, Berlin, 1997), pp. 513–525
17. H. Chen, S. Potluri, F. Koushanfar, BioChipWork: reverse engineering of microfluidic biochips, in *Proceedings of IEEE International Conference on Computer Design (Newton, MA)* (2017), pp. 9–16
18. Y. Moradi, M. Ibrahim, K. Chakrabarty, U. Schlichtmann, Fault-tolerant valve-based microfluidic routing fabric for droplet barcoding in single-cell analysis, in *2018 Design, Automation & Test in Europe Conference & Exhibition* (2018)
19. M. Mesbahi, State-dependent graphs, in *Proceedings of IEEE Conference on Decision and Control (Lahaina, HI)*, vol. 3 (2003), pp. 3058–3063
20. J. Tang, M. Ibrahim, K. Chakrabarty, R. Karri, Security implications of cyberphysical flow-based microfluidic biochips, in *Proceedings of IEEE Asian Test Symposium (Taipei)* (2017), pp. 110–115
21. J. Tang, M. Ibrahim, K. Chakrabarty, R. Karri, Security trade-offs in microfluidic routing fabrics, in *Proceedings of IEEE International Conference on Computer Design (Newton, MA)* (2017), pp. 25–32
22. S.-I. Minato, N. Ishiura, S. Yajima, Shared binary decision diagram with attributed edges for efficient Boolean function manipulation, in *Proceedings of IEEE/ACM Design Automation Conference* (1990), pp. 52–57
23. R.I. Bahar, E.A. Frohm, C.M. Gaona, G.D. Hachtel, E. Macii, A. Pardo, F. Somenzi, Algebraic decision diagrams and their applications. Formal Methods Syst. Des. **10**(2–3), 171–206 (1997)
24. D.E. Knuth, A.C. Yao, The complexity of non-uniform random number generation, in *Algorithms and Complexity: New Directions and Recent Results*, ed. by J.F. Traub (Academic, New York, 1976)
25. H. Zhou, H.-L. Chen, J. Bruck, Synthesis of stochastic flow networks. IEEE Trans. Comput. **63**(5), 1234–1247 (2014)
26. R.E. Bryant, Graph-based algorithms for Boolean function manipulation. IEEE Trans. Comput. **100**(8), 677–691 (1986)
27. K.M. Horsman, J.M. Bienvenue, K.R. Blasier, J.P. Landers, Forensic DNA analysis on microfluidic devices: a review. J. For. Sci. **52**(4), 784–799 (2007)
28. J. El-Ali, P.K. Sorger, K.F. Jensen, Cells on chips. Nature **442**(7101), 403–411 (2006)
29. S. Hosic, S.K. Murthy, A.N. Koppes, Microfluidic sample preparation for single cell analysis. Anal. Chem. **88**(1), 354–380 (2015)
30. A.M. Klein, L. Mazutis, I. Akartuna, N. Tallapragada, A. Veres, V. Li, L. Peshkin, D.A. Weitz, M.W. Kirschner, Droplet barcoding for single-cell transcriptomics applied to embryonic stem cells. Cell **161**(5), 1187–1201 (2015)
31. T.M. Cover, J.A. Thomas, *Elements of Information Theory* (Wiley, New Delhi, 2012)
32. M. Ibrahim, K. Chakrabarty, K. Scott, Synthesis of cyberphysical digital-microfluidic biochips for real-time quantitative analysis. IEEE Trans. Comput. Aided Des. Integr. Circuits Syst. **36**(5), 733–746 (2017)
33. J. Pearl, *Probabilistic Reasoning in Intelligent Systems: Networks of Plausible Inference* (Morgan Kaufmann, San Francisco, 1988)
34. L. Xing, S.V. Amari, *Binary Decision Diagrams and Extensions for System Reliability Analysis* (Scrivener, Beverly, 2015)
35. C. Clos, A study of non-blocking switching networks. Bell Labs Tech. J. **32**(2), 406–424 (1953)
36. J. Duato, S. Yalamanchili, L.M. Ni, *Interconnection Networks: An Engineering Approach* (Morgan Kaufmann, San Francisco, 2003)

Chapter 6
Conclusions

6.1 Security in Practice

This book presented a comprehensive overview of emerging security and trust issues in cyberphysical microfluidic systems, with a focus on countermeasures against actuation tampering attacks. Each of these countermeasures provided probabilistic guarantees of integrity. Although case studies were presented to illustrate their utility, there admittedly may still be some ambiguity as to the interpretation of the metrics. That is, do these countermeasures *really* make a microfluidic platform more secure? The following reasons point to the answer being a definitive "yes":

- *Cost to the attacker*: So far, we have not mentioned whether there is a cost to the attacker in attempting to tamper with a microfluidic platform. This is because the costs associated with carrying out an attack are inherently difficult to quantify. For instance, consider a company wishing to conduct corporate sabotage by causing repeated failures of a competitor's products. If the sabotage can be successfully attributed, there would be legal and economic repercussions.
- *Over-estimation*: We often analyze security in terms of worst-case scenarios. This likely over-estimates the impact. An attacker may not be savvy enough to attempt to penetrate the weakest parts of a defense. For instance, in the tamper-resistant pin-constrained DMFB, we eliminated several don't-cares, but some time-steps of an assay are more easily hacked than others. An attacker may not know or care to target these weaker time-steps. Also, attacks that occur repeatedly or in parallel will necessarily lower the probability that they will evade detection.
- *Deterrence*: The fact that countermeasures exist at all may be a deterrent to some would-be attackers. We assume a lot about an adversary—that they are devious and can steal information on how the defenses work and then try to bypass them. But the knowledge that a countermeasure exists can act as a deterrent (similarly as to traditional physical security [1]).

© Springer Nature Switzerland AG 2020
J. Tang et al., *Secure and Trustworthy Cyberphysical Microfluidic Biochips*,
https://doi.org/10.1007/978-3-030-18163-5_6

6.2   Future Work

Here we present some potentially fruitful areas for future research.

6.2.1   Intellectual Property Protection

Intellectual property (IP) protection will likely be the most fruitful topic for future researchers. Corporations have an interest in preserving their IP, and currently, few techniques have been described that are specific to microfluidic technologies. Furthermore, there is a need for IP protection at the *fluidic* level. This will be challenging since the chemical domain is difficult to model. Currently existing techniques include the use of a "fluidic multiplexer" for bioassay encryption [2], bioassay locking based on the insertion of dummy mix-split operations [3], and a digital rights management scheme based on PUFs [4]. The simplest way forward would be to adapt existing IP protection concepts to microfluidics. Watermarking, obfuscation, and metering have all been identified as potential solutions [5], though care must be taken to address the microfluidic aspects of the IP as much as possible.

6.2.2   Micro-Electrode-Dot-Array Biochips

Micro-electrode-dot-array (MEDA) biochips are the next step in the evolution of digital microfluidics technology [6, 7]. Instead of a coarse-grained array of electrodes where each electrode roughly corresponds to a single droplet, MEDA biochips are fine-grained. A single droplet can occupy multiple electrodes. This is attractive as it gives more degrees of freedom for droplet manipulation; droplets can now move diagonally, and be split into non-uniform sizes. Already, several publications have developed EDA techniques for MEDA biochips [8–11], and it will be interesting to see what the advantages and disadvantages of this technology are in terms of security. One important thing to note is that MEDA biochips are often fabricated using semiconductor processes, therefore allowing the integration of sensors and logic within the biochip. This increases the sensing capability, but also leads to the possibility of hardware Trojan insertion.

6.2.3   Trusted Sensing with SensorPUFs

Sensing is a fundamental operation on a CPMB. All of the complex protocols that are developed for these platforms are often used just to ascertain a certain property of a sample-under-test. Ensuring the integrity of this sensor reading is critical, but could be difficult if the microfluidic platform is physically compromised.

Sensor physical unclonable functions (SensorPUFs) are sensors that are designed to leverage natural manufacturing variations [12] for security purposes. A challenge-response protocol is used to query the SensorPUF. The mapping between challenge bits and response bits is a function of the physical quantity that the sensor is designed to measure, and this mapping is randomly determined for each fabricated SensorPUF. The SensorPUF must be enrolled by a trusted party before it is deployed in the field. When a party wishes to query the sensor, it issues a challenge and then looks up the response in its database of challenge-response pairs. This pairing implicitly gives the sensor reading.

A SensorPUF is one example of a virtual proof of reality [13]. Few SensorPUF designs have been proposed: a pressure sensor based on MEMS relays [14] and a coated optical sensor [12]. Challenge-response pairs (CRPs) must be stable across operating conditions not related to the physical quantity being measured. Ensuring that there are enough usable CRPs for the lifetime of the device is challenging in practice. Additionally, certain PUF architectures have been shown to have weaknesses against model building attacks [15]. One key point about SensorPUFs used in biochips is that biochips are often designed to be one-time use, in order to prevent contamination of samples. This provides an excellent opportunity to leverage "weak" PUFs [15], as there is no need to repeatedly make measurements.

6.2.4 Self-Erasure and Self-Destruction

Physical security of a device after its useful lifetime is not guaranteed. While it is an open question as to how much information an attacker can collect from a used biochip, it is expected that fluid residues on the chip surface, as well as waste droplets collected in reservoirs, can be used to glean sensitive information. A biochip that is programmed to self-destruct or self-clean after its useful life would prevent such attacks.

In general-purpose DMFBs, wash droplets are often utilized to clean electrodes and prevent surface contamination [16]. These operations can be automatically generated by the synthesis toolchain to erase electrodes occupied by fluids containing sensitive information. Alternately, the sensitive information could be destroyed by introducing a reagent that is known to destroy or denature the samples on the biochip. Actuators could also be used, for instance, to overheat DNA samples to the point of denaturation.

In some cases, it may be beneficial for the microfluidic device to include a mechanism for self-destruction in the event of an invasive attack. For instance, a biochip that is designed to store samples for long-term processing and observation should be able to prevent an attacker from extracting the contents of the biochip. MEMS sensors sensitized to the physical characteristics of an invasive attack could both initiate a self-destruction sequence and power the mechanism used to self-destruct [17, 18].

6.2.5   Experimental Demonstration

Nearly all of the work to date in CPMB security has been theoretical in nature. Experimental demonstrations are needed to create threat models that are more in line with reality and to better evangelize the field. Simple experiments could involve replication of many of the VLSI-based threats in microfluidics, such as hardware Trojans. More ambitious experiments would focus more on biochemistry and fluid-level problems. Ideally, work should be done to incorporate new security measures into a commercial offering.

6.3   Summary

We summarize the contributions of this work as follows:

- Security and trust issues in cyberphysical microfluidic systems were analyzed and categorized. The taxonomy derived in this work will provide a solid point of reference for future work, while the analysis itself can be mined for ideas.
- The tamper-resistance property of pin-mapped DMFB designs was defined and analyzed, and a new tamper-resistant pin mapping algorithm was developed. To further improve this property, the concept of an indicator droplet was introduced and its routing was solved with a sliding window ILP-based algorithm.
- Hardware used in error recovery for cyberphysical DMFBs were leveraged for security purposes as a randomized checkpoint system. The theoretical performance of the system was derived and improved through graph-based strategic static checkpoint placement.
- Analysis frameworks for transposer-based routing fabrics were developed under both fault injection and random attack models. A graph-based probability distribution synthesis methodology was derived and optimally solved using ILP and an $O(N^2)$ algorithm. Simple constructions were provided for ensuring fail-safe and uncontrollable behavior under random attack.

We also reiterate some of the major themes of this book:

- Microfluidics is a technology that has significant security and trust implications, as well as opportunities due to their unique properties such as reduced complexity and longer time scales.
- Security and trust may be the attributes that finally push microfluidic technologies into the mainstream.
- Designers are well advised not to wait for insecure standards to be ingrained into the technology; even commercial IoT devices have been shown to be insecure, despite years of warnings from academic researchers.
- Microfluidic security is not a replacement for traditional computer and cyber-physical system security; instead, it should be thought of as either complementing or augmenting them. We achieve this through design focused on low-level hardware.

We close by noting that it is far too easy to be skeptical about speculative work in an emerging technology. We believe such a viewpoint is shortsighted and hope that this book has made progress in disabusing readers from such beliefs. This book features some of the earliest work in what we anticipate will become a burgeoning field. We look forward to seeing other researchers taking the ideas presented here and running with them.

References

1. N.V. Lasky, B.S. Fisher, S. Jacques, 'Thinking thief' in the crime prevention arms race: lessons learned from shoplifters. Secur. J. **30**(3), 772–792 (2017)
2. S.S. Ali, M. Ibrahim, O. Sinanoglu, K. Chakrabarty, R. Karri, Microfluidic encryption of on-chip biochemical assays, in *Proceedings of IEEE Biomedical Circuits and Systems Conference (Shanghai)* (2016), pp. 152–155
3. S. Bhattacharjee, J. Tang, M. Ibrahim, K. Chakrabarty, R. Karri, Locking of biochemical assays for digital microfluidic biochips, in *Proceedings of IEEE European Test Symposium (Bremen)* (2018), pp. 1–6
4. C.-W. Hsieh, Z. Li, T.-Y. Ho, Piracy prevention of digital microfluidic biochips, in *Proceedings of Asia South Pacific Design Automation Conference (Chiba)* (2017), pp. 512–517
5. S.S. Ali, M. Ibrahim, J. Rajendran, O. Sinanoglu, K. Chakrabarty, Supply-chain security of digital microfluidic biochips. Computer **49**(8), 36–43 (2016)
6. G. Wang, D. Teng, Y.-T. Lai, Y.-W. Lu, Y. Ho, C.-Y. Lee, Field-programmable lab-on-a-chip based on microelectrode dot array architecture. IET Nanobiotechnol. **8**(3), 163–171 (2013)
7. K.Y.-T. Lai, Y.-T. Yang, C.-Y. Lee, An intelligent digital microfluidic processor for biomedical detection. J. Signal Process. Syst. **78**(1), 85–93 (2015)
8. Z. Li, K.Y.-T. Lai, P.-H. Yu, T.-Y. Ho, K. Chakrabarty, C.-Y. Lee, High-level synthesis for micro-electrode-dot-array digital microfluidic biochips, in *Proceedings IEEE/ACM Design Automation Conference* (2016), p. 146
9. O. Keszocze, Z. Li, A. Grimmer, R. Wille, K. Chakrabarty, R. Drechsler, Exact routing for micro-electrode-dot-array digital microfluidic biochips, in *Proceedings of Asia South Pacific Design Automation Conference* (2017), pp. 708–713
10. Z. Li, K.Y.-T. Lai, P.-H. Yu, K. Chakrabarty, T.-Y. Ho, C.-Y. Lee, Built-in self-test for micro-electrode-dot-array digital microfluidic biochips, in *Proceedings of IEEE International Test Conference* (2016), pp. 1–10
11. Z. Li, K.Y.-T. Lai, P.-H. Yu, K. Chakrabarty, M. Pajic, T.-Y. Ho, C.-Y. Lee, Error recovery in a micro-electrode-dot-array digital microfluidic biochip? in *Proceedings of IEEE/ACM International Conference on Computer-Aided Design* (2016), p. 105
12. K. Rosenfeld, E. Gavas, R. Karri, Sensor physical unclonable functions, in *Proceedings of International Symposium on Hardware Oriented Security and Trust* (2010), pp. 112–117
13. U. Ruhrmair, J. Martinez-Hurtado, X. Xu, C. Kraeh, C. Hilgers, D. Kononchuk, J.J. Finley, W.P. Burleson, Virtual proofs of reality and their physical implementation, in *Proceedings of IEEE Symposium on Security & Privacy* (2015), pp. 70–85
14. J. Tang, R. Karri, J. Rajendran, Securing pressure measurements using SensorPUFs, in *Proceedings of IEEE International Symposium on Circuits and Systems (Montreal)* (2016)
15. C. Herder, M.-D. Yu, F. Koushanfar, S. Devadas, Physical unclonable functions and applications: a tutorial. Proc. IEEE **102**(8), 1126–1141 (2014)
16. Y. Zhao, K. Chakrabarty, Cross-contamination avoidance for droplet routing in digital microfluidic biochips. IEEE Trans. Comput. Aided Des. Integr. Circuits Syst. **316**, 817–830 (2012)

17. D. Shahrjerdi, J. Rajendran, S. Garg, F. Koushanfar, R. Karri, Shielding and securing integrated circuits with sensors, in *Proceedings of IEEE/ACM International Conference on Computer-Aided Design* (IEEE Press, Piscataway, 2014), pp. 170–174
18. R.A. Coutu, S.A. Ostrow, Microelectromechanical systems (MEMS) resistive heaters as circuit protection devices. IEEE Trans. Compon. Packag. Manuf. Technol. **3**(12), 2174–2179 (2013)

Index

© Springer Nature Switzerland AG 2020
J. Tang et al., *Secure and Trustworthy Cyberphysical Microfluidic Biochips*,
https://doi.org/10.1007/978-3-030-18163-5

Printed in the United States
By Bookmasters